S0-AQI-696

Value Theory *and* Education

Edited by

Peter F. Carbone, Jr.
Duke University

ROBERT E. KRIEGER PUBLISHING COMPANY
MALABAR, FLORIDA
1987

Original Edition 1987

Printed and Published by
ROBERT E. KRIEGER PUBLISHING COMPANY, INC.
KRIEGER DRIVE
MALABAR, FLORIDA 32950

Printed in the United States of America.

Library of Congress Cataloging-in-Publication Data

Value theory and education.

Bibliography: p.
Includes Index.
1. Moral education—United States. 2. Values.
I. Carbone, Peter F.
LC311.V33 1987 370.11'4 86-27822
ISBN 0-89874-976-X
10 9 8 7 6 5 4 3 2

In memory of my
mother and father

Contents

Preface

The aim of this book is to demonstrate the many-sided relationship between value theory and educational policy and practice. Although one aspect of that relationship—the connection between moral philosophy and moral education—has been thoroughly examined during the past three decades, relatively little has been written on the general import of value theory for the educational process as a whole. Consequently, the present volume seeks to explore the broader implications of the relationship in question, without losing sight of the continuing debate on moral education.

Although professional philosophers and philosophers of education may find the book a useful resource for teaching or review purposes, the collection was assembled with a wider audience of educators, students, and laypersons in mind, an audience with an informed, general interest, perhaps, rather than a professional competence in the area of ethics and education. Hence the selected readings were intended to comprise a representative sample of some of the more thoughtful, yet lucid writings on the subject, many of them the contributions of respected authorities in the field.

The articles include a wide range of topics and extend over several decades. Some are of recent vintage, but date of publication was accorded less weight in the selection process than intrinsic value, clarity and force of argument, and effective elucidation of the issues considered in the six sections. Moreover, one of the objectives of the collection is to furnish some historical perspective for evaluating current issues regarding values and education.

The book will have served its purpose if it clarifies to some extent the complex relationships between ethics and education, if it helps to link the practical concerns of educators to the theoretical conclusions of ethical theorists, and if it encourages additional efforts along these lines.

Acknowledgments

I am grateful to the authors and publishers for their cooperation.

I also wish to express my gratitude to Mary Roberts of the Krieger Publishing Company for editorial suggestions and to Agnes Barnhardt for help in preparing the manuscript for publication.

Finally, I want to thank my wife, Claire, for proofreading assistance and, appropriately enough, given the nature of the subject matter, for moral support during the lengthy preparation of the book.

Introduction

I

During the past thirty years American educators have become increasingly interested in the concept of "moral education" and, to a somewhat lesser extent, in the application of value theory to educational problems and concerns. This recent surge of interest represents something of a departure from the attitudes of the previous generation of educators, most of whom seemed content to place major responsibility for moral education with the home and church.

It would be inaccurate, however, to interpret the contemporary interest in moral education and related areas as a sharp break with tradition. Rather, it should be viewed in historical perspective as the reappearance of a strain long evident in Western educational history. As a matter of fact, most of the major educational philosophers from Plato to Dewey have been at least as concerned with building character and imparting values as with cognitive development and the dissemination of factual information. Indeed, some of the leading proponents of cognition and rationality—Aristotle, Kant, and Dewey among them—actually devoted much of their educational philosophizing to ethical issues such as the problem of deciding just which moral dispositions or character traits education should foster. One finds similar concerns expressed in the works of Plato, Rousseau, Pestalozzi, Herbart, Froebel, and any number of other thinkers whose frequent appearances in standard anthologies attest to their lasting influence on educational philosophy.

This emphasis on moral education or character development is one indication that leading educational theorists have always recognized what R. S. Peters calls "the normative aspect of education." "Education," according to Peters,

> implies that something worth while is being or has been intentionally transmitted in a morally acceptable manner. It would be a logical contradiction to say that a man had been educated but that he had in no way changed for the better, or that in educating his son a man was attempting nothing that was worth while.[1]

Accordingly, school systems are established with the fond hope that on the whole the educational process will bestow more good than harm on children, that the young scholars will emerge from the educational tunnel somewhat improved for having taken the trip. We even dare to presume, moreover, that some of the benefits derived by individuals from formal education will eventually accrue to society at large.

Schools are established, then, for the purpose of fostering that which society considers worthwhile. Hence the observation that all educational systems reflect prevailing social and cultural values has become commonplace. Our own system is perhaps as good an example of this as any because of the strong flavor of moral zeal discernible throughout much of our social and intellectual history. Henry May summarizes the American preoccupation with morality as follows:

> A history of moralism would come close to being a history of American thought. For Jonathan Edwards and his heirs, the morality of the universe had depended on the exertion of God's will, and had demanded the damnation of most men. As this gloomy view succumbed to the European enlightenment and the American environment, universal morality itself had grown no less secure. For Jefferson, for instance, it had been a matter of self-evident natural law, confirmed by man's innate moral sense. For Emerson the moral meaning of the universe had been something that all men could find by looking either inward or outward through the eyes of the spirit. For the drier academic pundits of the early nineteenth century, morality had been demonstrable through logic; it was necessary to thought, we could not help believing it, so it must be true. Materialism and skepticism were the discredited notions of a few heathen Greeks and dissolute Frenchmen.[2]

All of this has been clearly reflected in our educational history. One of the first orders of business for the Massachusetts Puritans, for example, was to establish educational institutions in order to preserve their norms and values as well as to train clergymen. To those ends they founded the Boston Latin School and Harvard College, and required the establishment of elementary schools in towns of fifty families or more, plus a secondary (grammar) school in towns of at least one hundred families, all within thirty years of the settling of Plymouth Colony.

Puritan values were rooted in religious convictions, of course, and these convictions were evident in the makeup of our early schools. Although the gradual secularization of American society reduced the influence of the churches on schools, public education has nevertheless retained much of its normative character. The common school movement of the mid-nineteenth century, for example, was inspired to a considerable degree by Horace Mann's vision of an educational system that would instill a common core of values, acceptable to the various segments of a diverse society.

Later in the century, during the period of heavy immigration, the schools were expected to "Americanize" the immigrants by initiating them into our system of civic, cultural, and social values. Thus, social cohesion and civic responsibility would be maintained in a society becoming increasingly pluralistic.

The idea that the moral influence of the school could extend to the home and neighborhood, and ultimately find its way into the national character carried over into the first half of the twentieth century, notably in the work of John Dewey and his followers, who thought the school could, and should, be employed as an agent of social betterment. By that time, however, sweeping social change fueled by rapid urban growth, the legacy of disillusionment left by World War I, and the destabilizing impact of a series of intellectual sea changes had shaken the moral foundations of American practical idealism,[3] and fostered a trendy subjectivism that later proved inimical to the notion of moral education.

The transformation of much of rural agrarian America into sprawling urban complexes, for example, had removed the restraints of small-town conformity and loosened the bonds of the extended family. And in the process, traditional attitudes, values, and patterns of behavior were modified.

The rapid and haphazard growth of the city also called into question the traditional American faith in progress. Faced with the harsh realities of overcrowding, poverty, crime, vice, and corruption, many Americans were forced to reconsider the conventional view that change in the United States, as if aided by supernatural intervention, invariably signaled progress toward a higher plane of social and cultural existence.[4] As Richard Hofstadter pointed out:

> The whole cast of American thinking in this period was deeply affected by the experience of the rural mind confronted with the phenomena of urban life, its crowding, poverty, crime, corruption, impersonality, and ethnic chaos. To the rural migrant raised in respectable quietude and the high-toned moral imperatives of evangelical Protestantism, the city seemed not merely a new social form or way of life but a strange threat to civilization itself.[5]

The war that neither ended all wars nor made the world safe for democracy added dramatically to the disillusionment. "It destroyed faith in progress," William Leuchtenburg writes, "but it did more than that—it made clear to perceptive thinkers that they had misread the Progressive era and the long Victorian reign of peace, that violence prowled underneath man's apparent harmony and rationality."[6] Americans who once viewed themselves as part of an essentially moral universe where truth, justice, honesty, decency, and human dignity either prevailed or were attainable were now hearing from writers such as Henry James, Ernest Hemingway,

Ezra Pound, and T. S. Eliot that the immediate past had been a "fool's paradise," that many of them were part of a "lost generation," and that all of them were caught up in a "botched civilization," residing in a "wasteland" without value or purpose. Of course many Americans had drawn similar conclusions without benefit of tutoring from intellectuals. Thus Leuchtenburg is undoubtedly correct in arguing that:

> Intellectuals led the assault on the groups which had traditionally exercised moral authority, but the "revolution in morals" would have taken place without them. The nation had lost its fear of the wrath of God and its faith in the nineteenth-century moral standards the churches had supported. It no longer had the same reverence for the old mores, ... [7]

Yet the fact remains that the nineteenth-century moral consensus proved highly vulnerable to events at least partly because it had been on the defensive since the middle of the century, its main tenets rendered doubtful by a series of scientific, socio-economic, psychological, and philosophical ideas that had gradually filtered into the public consciousness. It might be instructive to touch briefly on a few of these ideas, with emphasis on those that coalesced to form an intellectual climate clearly antithetical to the pursuit of moral education.

II

The intellectual inroads began with a "materialistic," "mechanistic" interpretation of man and his world that was diametrically opposed to the earlier nineteenth-century moral outlook. As May has observed:

> The discoveries of Darwin's generation challenged far more than orthodox Christianity; they inescapably suggested the distressing outlines of a mechanistic universe. Not only man himself, but life as a whole, even the existence of the planet and the solar system were apparently to be explained by chemical-physical processes which had little to do with traditional ideas of purpose or destiny . . . . A deadly fear of mechanistic materialism lay in back of the intense doubts and hard-won affirmations of the great Victorian sages.[8]

"Dialectical materialism" with its insistence on the existence of a causal relationship between social and cultural conditions on the one hand and prevailing economic relations on the other, seemed equally detrimental to the notion of man as autonomous moral agent. For many, the Marxist promise of a better future was more than offset by its core of historical determinism, which was perceived to be at marked variance to traditional ideas of human freedom and dignity.

To many observers, what little remained of that freedom and dignity following the onslaughts of mechanism, Darwinism, and Marxism underwent further buffeting at the hands of contemporary psychological theory. The Freudian suggestion that it was man's unconscious drives and impulses rather than sweet reason that regulated much of his conduct, and the behaviorist view that human behavior was largely a function of environmental conditioning were equally unsettling to those steeped in traditional humanistic values. In neither case did it seem appropriate any longer to praise or blame people for their actions. Rather, it became more fashionable to account for human behavior in terms of subconscious drives or external stimuli.

Goldman described the then prevailing climate of opinion as follows:

> "It's all relative," the advanced thinkers chorused, and so it was. Human attributes? Obviously a product of environment. The actions of human beings? Clearly the result of economic and psychical compulsions. Political or religious ideas? Naturally the reflection of property relationships and of mass psychology. Everything was "explained" or "understood" in the bright light of science . . . . Here was freedom from all absolutes, from all codes, ...[9]

Thus, in somewhat paradoxical fashion a set of more or less deterministic doctrines, each of them laying claim (some more convincingly than others, perhaps) to varying degrees of scientific objectivity, had been transmuted into attitudes of relativism and subjectivism among intellectuals and, presumably, at least to some extent among the general public as well. But for absolutist holdouts there was yet more to endure, for even the arcane achievements of early twentieth-century science served to reinforce the growing skepticism. To be sure, the "new physics," as Crane Brinton noted, "helped put the finishing touch on the destruction of the simple nineteenth-century notions of scientific causation, notions that conceived all relations in the universe on a neat mechanical model, ..."[10] Yet, the cluster of scientific and philosophical ideas that supplanted the "mechanical model," which had posed so serious a threat to nineteenth-century moralism, did nothing to restore the vitality of that earlier point of view. On the contrary, esoteric new insights, such as those contained in the theory of relativity, for instance, had the opposite effect. To illustrate, although Einstein himself probably cringed at the suggestion, he became in the public mind, as Brinton noted,

> the man who stood for relativity, for the notion that things look different to

> observers at different places and different times, that truth depends on the point of view of the seeker for truth, that a man moving at one rate of speed sees everything quite differently from a man moving at another rate, that, in short, there is no absolute Truth, but only relative truths.[11]

If, as Brinton suggested, the new physics applied the finishing touch to the mechanistic outlook, the beginning touches, so to speak, had been provided by some late nineteenth and early twentieth-century thinkers, William James and John Dewey notable among them, whose intellectual orientation was naturalistic (with special emphasis on the evolutional perspective) but who rejected the more extreme forms of naturalism, that is, the materialistic or mechanistic doctrines that had depicted the universe as closed, fixed, and machinelike; or human beings as for the most part regulated, controlled, or determined by environmental forces. To naturalists of the James-Dewey persuasion, the universe was open-ended, dynamic, uncertain, and therefore somewhat precarious. The world they looked out upon was one of constantly emerging novelty, hence a world in which knowledge and values were temporal and contingent upon changing conditions.[12]

The relativism yielded by this version of naturalism (unlike that derived from the deterministic views summarized earlier) was not at all paradoxical. Indeed, it was an intentional, integral component in the theory, and it was equally apparent in much of the theoretical social science carried on at the time.[13] Thus turn-of-the-century social thought, despite its opposition to the fixities and constraints of various forms of materialism, could hardly be viewed as a rescue operation for a beleagured moral universalism. Instead, these newer styles of thought represented yet another assault on the traditional faith in the certainty and timelessness of moral convictions.

III

The possibility of arriving at absolute truth even with regard to factual matters, much less in the realm of values, has also been discounted in three of the major philosophical perspectives of our time: pragmatism, existentialism, and logical positivism. For despite the many differences within and between these schools of thought, their adherents have shared an aversion to epistemological and ethical absolutism, an aversion that has increasingly called into question the possibility of objectivity in ethics and, by extension, in values education. Pragmatism, for example, (with strong ties through James and Dewey to naturalism) denied that truth was a fixed property of ideas. Rather, an idea was true insofar as it helped us to relate satisfactorily to the various elements of our experience, though what constituted satisfaction was a point of contention among leading pragmatic thinkers. (James, for example, was comfortable with considerably more subjectivity than either Charles Sanders Peirce or Dewey was willing to accept.) In any event, the notion was that ideas were true if they worked, that is, if they helped us to cope in a problematical existence in which very little was certain or unchanging. Clearly, one should be prepared in such a world to revise or discard attitudes, beliefs, or values in light of altered conditions.

The open-endedness so characteristic of pragmatism was also one of the defining traits of existentialism and served as a connecting link to the former in spite of important differences between the two philosophies. Existentialists are a diverse lot and can often be found in disagreement with one another on various points, but like pragmatists they tend to share a healthy skepticism concerning the likelihood of knowledge that is certain or values that are permanent. Indeed, with their insistence on the arbitrariness of social norms and standards, their celebration of personal freedom and decision making, and their denial of reliable guides to conduct, they often carry ethical subjectivity to even greater lengths than do the pragmatists.

As a result of the influence of John Dewey and his colleagues, pragmatism has had considerable impact on educational theory; and many of the leading tenets of both pragmatism and existentialism have, through extensive popular appeal, contributed significantly to the shaping of contemporary thought. Logical positivism, on the other hand, has never achieved so high a level of public popularity, but in lending support to the emotive theory of ethics, it may have exerted even more influence on recent moral philosophy than either of the other two systems. By counting as meaningful only statements that could be either verified or falsified at least in principle, the positivists, in effect, relegated propositions in fields such as metaphysics, religion, and ethics to the realm of the meaningless or nonsensical. Alternatively, if value judgments carry any meaning at all, so the argument ran, it is not information-bearing, cognitive meaning; rather, it is "emotive" meaning, which merely reveals the speaker's feelings. On this account, therefore, value judgments (including educational values of course) cannot be shown to be true or false. Furthermore, if this view is correct, it would seem to follow that moral education is virtually indistinguishable from indoctrination.

As a result, logical positivism contributed significantly to contemporary ethical relativism and thereby reinforced prevailing misgivings regarding the legitimacy of moral, or values, education. At the same time, however, so radical a departure from traditional "meta-ethical" theory all but assured some sort of philosophical "backlash" and an early arrival on the scene of a less drastic point of view. As Douglas Sloan has observed:

> By baldly denying the rationality of ethical statements the positivists and emotivists had laid down a challenge and set the agenda of subsequent ethical inquiry. If the emotivists were not correct, how and in what way could ethical judgments in fact be said to have a rationality and integrity in their own right?[14]

The reply to that question, furnished by a newly emerging group of moderate "noncognitivist" moral philosophers, amounted to a turning point of sorts, for it led eventually to an intellectual climate more receptive to both normative ethics and moral education.

IV

The emotivist thesis had been a reaction against the assumption that ethical terms referred either to natural phenomena such as interests or desires (the naturalists' view) or to "non-natural" properties such as goodness or rightness (the position taken by members of the intuitionist school of ethics, which dominated early twentieth-century moral philosophy in England). According to the emotivists both premises were false; ethical terms were "non-referential" or "noncognitive." They functioned not as referents to properties, whether natural or non-natural, but rather as emotive expressions of attitudes or disguised imperatives. As might be expected, the "anything-goes," "to-each-his-own" implications of this view were repugnant to many moral philosophers and in due course elicited a negative reaction from the more moderate school of "noncognitivism" that gained ascendency in the post World War II period.[15]

Like the emotivists, the moderate noncognitivists rejected ethical naturalism and intuitionism, but they also denied that ethical propositions could be reduced to emotional utterances, commands, or mere expressions of taste. Though value judgments were neither empirically verifiable nor logically demonstrable, they were nevertheless rationally justifiable in the sense that the most defensible judgments were supported by the best reasons. When one engages in ethical discourse, in other words, the appeal is to reason, not to arbitrary personal preferences.

Since most of the moderate noncognitivists were nurtured in contemporary linguistic analysis, they were more attentive to meta-ethical questions, that is, to problems relating to the meaning and justification of ethical terms and propositions than to normative judgments of obligation, or claims as to what is right or wrong, good or bad, valuable or worthless. All the same, some prominent analysts began to explore various connections between ethics and education, including, in some cases, issues relating directly to moral education.[16] These writers concerned themselves with the meaning and structure of ethical discourse, to be sure, but they also dealt, at least theoretically, with the acquisition of values and other substantive educational issues.

Gradually these concerns found their way into the educational literature and into syllabi for courses in philosophy of education, and by the early 1960s moral education was becoming a topic of interest to professional educators at the precollege level as well as to ethical theorists and educational philosophers. Then, over the next few years, interest in the subject was given added impetus by a series of events that led increasing numbers of Americans to question the nation's sense of values and the wisdom of having neglected moral education in the nation's schools. Assessing the effects of these events in 1975, Purpel and Ryan wrote:

> The last dozen years in this country have brought one jarring event after another: a wave of political assassinations, the violent rupture of the black civil rights struggles, the student revolution that polarized young and old, the Vietnam war, and finally, the agony of Watergate. We are numb from the impact of these events. We have lost faith in many of our institutions.[17]

To some observers these events, coupled with soaring crime and divorce rates, the "revolution" in sexual attitudes, and pervasive rejection of various kinds of authority, suggested a process of moral decay so pandemic as to be nearing crisis proportions.

All of this contributed to a willingness to re-evaluate moral relativism, at least in its more extreme forms, and to view the school as the obvious place in which to initiate corrective educational measures. The result, during the 1970s, was a spate of books and articles on moral education, a full-blown "values clarification" movement,[18] and a number of courses and instructional programs, many of them inspired by Lawrence Kohlberg's research on moral development in children. And there has been no discernible waning of interest in the 1980s. At the annual meetings of the national Philosophy of Education Society (as well as those of its regional affiliates), for example, Kohlberg's views continue to inspire spirited discussion, and Carol Gilligan's research on the relationship between gender and moral outlook has added a new and controversial dimension to the continuing debate.[19] Meanwhile, the values clarification movement continues to gain in popularity, and the new right, which often tends to equate morality with religion, has entered the debate on moral education with an agenda that includes school prayer, equal time for creationism in the science curriculum, and constant vigilance against the appearance in our public schools of what conservatives in some quarters refer to as "secular humanism."[20]

We have moved, then, in a remarkably short time, from widespread neglect of formal moral education to an enthusiasm for the subject that at times seems to have reached fad proportions. It is interesting to note, however, that in comparison to the ample output on moral education, the literature addressed to the general import of axiology for educational concerns has been rather sparse.[21] Moreover, one seldom finds even a general introduction to value theory and its implications for education among the course requirements for teachers and other educational professionals.

The oversight is rather curious in view of the fact that the educational process is fairly saturated with ethical or value considerations. It is certainly fitting that moral education should emerge as the prime example of the close relationship between ethics and education; since teachers cannot avoid imparting values in one way or another in the normal course of their activities *qua* teachers, moral education in some sense is unavoidable. Thus the real question is not whether, but how it should be carried on. But moral education is simply the most visible manifestation of this multifaceted relationship. It is quite clear, for example, that any serious discussion of educational policy leads quickly to a consideration of educational aims, which are based on assumptions concerning the good life, the good society, and any number of additional ethical considerations. Education, after all, is a purposeful activity, one that inevitably incorporates value judgments of various kinds. What we consider "good," "right," or "important" constantly guides our practice, whether consciously or not.

Since schools are expected to achieve social purposes while they educate, ethical concepts also come into play at the boundaries between social philosophy and educational policy, where issues such as integration, student rights, equality of opportunity, and local control are debated. Disputes over public policy that affect our schools always entail certain assumptions about the meaning of concepts such as "justice," "equality," and "freedom" along with the economic, political, and psychological factors that usually receive greater attention in the media.

One could go on in this vein, pointing to examples of how value judgments influence the instructional process in ways ranging from the basic learning climate of the classroom (stretching from permissive to authoritarian, for example) to the everyday tasks of selecting teaching materials or grading essays. Perhaps enough has been said, however, to indicate the extent to which issues pertaining to the nature of value judgments, the linguistic functions they perform, how or whether they can be justified, and so forth are of practical interest to educators at all levels.

None of this is intended to suggest that the relationship between moral philosophy and moral education is one of logical implication, or that educational values can be derived directly from axiology. There is simply no royal deductive road leading from ethical theory to educational values and principles of moral education. Thus it is unrealistic to expect a study of ethics to yield a formula for devising lists of educational aims, policy statements, dispositions to be fostered, or curricular priorities. On the other hand, a degree of familiarity with ethical theory is a prerequisite for making informed judgments in these areas. Philosophical insights applied to education cannot solve practical educational problems, but they can furnish the kind of perspective needed to deal intelligently with those problems.

The recent interest in moral education on the part of both moral and educational philosophers has clearly led to noticeable progress in illuminating and clarifying the issues and problems encountered by practitioners in the field. Since values education in the broader sense alluded to above has not yet received comparable attention, this anthology seeks to illustrate the relevance of value theory not only for moral education, but for other educational concerns as well.

The book will be divided into six sections as follows: The first section will provide an overview of moral philosophy and its implications for educational policy and practice in general as well as for moral education in particular. The purpose of this section is to furnish some historical and theoretical perspective from which to consider the relevance of ethical theory for all facets of the educational process. The next two sections will present discussions devoted to the establishment of educational aims and the setting of educational policy. These two sections will include some consideration of how social values such as freedom, equality, and democracy affect educational means and ends. The fourth section will explore a few of the ways in which value judgments enter into decisions made with regard to teaching and the curriculum. And finally, the fifth and sixth sections will be concerned with some of the central questions and concepts to be found in the continuing debate over how moral education should be construed and carried out. It is hoped that the overall organization of the collection will reveal some integration and continuity within and between sections.

NOTES

1. R. S. Peters, *Ethics and Education*, 2d ed. (London: George Allen & Unwin Ltd., 1970), p. 25.
2. Henry F. May, *The End of American Innocence: A Study of the First Years of Our Own Time* (Chicago: Quadrangle Books, 1964), p. 10.
3. See May, ibid., Chap. 2 for an informative discussion of the elements of practical idealism.

4. T. Harry Williams, Richard N. Current, and Frank Freidel, *A History of the United States Since 1865*, 2d ed., rev. (New York: Alfred A. Knopf, 1970), pp. 97–98.
5. Richard Hofstadter, *The Age of Reform: From Bryan to F. D. R.* (New York: Vintage Books, 1955), p. 176.
6. William E. Leuchtenburg, *The Perils of Prosperity, 1914–32* (Chicago: The University of Chicago Press, 1958), p. 142.
7. Ibid., p. 7.
8. May, *American Innocence*, p. 10.
9. Eric F. Goldman, *Rendezvous with Destiny: A History of Modern American Reform* (New York: Vintage Books, 1956), p. 241.
10. Crane Brinton, *The Shaping of the Modern Mind* (New York: Mentor Books, 1953), p. 218.
11. Ibid., p. 217.
12. See May, *American Innocence*, pp. 140–153, for a brief but instructive account of the James-Dewey version of naturalism.
13. Ibid., pp. 153–165.
14. Douglas Sloan, "The Teaching of Ethics in the American Undergraduate Curriculum, 1876–1976," *Education and Values*, ed. Douglas Sloan (New York: Teachers College Press, 1980), p. 230.
15. For an illuminating discussion of recent meta-ethical theories, see William K. Frankena, *Ethics* (Englewood Cliffs, N. J.: Prentice-Hall, 1963), Chapter 6 (reprinted in this volume).
16. The contributions of R. M. Hare, Paul Hirst, R. S. Peters, and John Wilson in England and Henry Aiken, William Frankena, and Israel Scheffler in the United States were particularly important in this respect.
17. David Purpel and Kevin Ryan, "Moral Education: Where Sages Fear to Tread," *Phi Delta Kappan* 56 (June 1975): 660.
18. It should be noted, however, that critics of the values-clarification point-of-view consider it overly relativistic. In this connection, see John S. Stewart, "Clarifying Values Clarification: A Critique," *Phi Delta Kappan* 56 (June 1975): 684–688 (reprinted in this volume).
19. See Carol Gilligan, *In a Different Voice* (Cambridge: Harvard University Press, 1983).
20. In this connection see Ben Brodinsky, "The New Right: The Movement and Its Impact," *Phi Delta Kappan* 64 (October 1982): 87–94. This issue of the *Kappan* also contains several responses to Brodinsky's article.
21. However, Philip G. Smith, ed., *Theories of Value and Problems of Education* (Urbana: University of Illinois Press, 1970) covers a wider range of topics than most published works in the field.

Part One: Ethical Theory

Introduction

In order to deal intelligently with unavoidable value judgments, educators need some general knowledge of ethical theory. In Part One, which includes two essays by William K. Frankena and one by R. S. Peters, we shall briefly explore the nature of ethics from a historical as well as a philosophical point of view.

Professor Frankena's first selection may be described as an attempt to "map" the logical terrain in the field of ethics. Having defined ethics or moral philosophy as "that study or discipline which concerns itself with judgments of approval and disapproval . . . rightness or wrongness, goodness or badness," Frankena divides ethics into two parts, "the theory of value or axiology" and "the theory of obligation." He then points to questions related to the meaning of ethical terms such as "good" or "right," and questions concerning the things that are good or the actions that are right, as some of the key issues in moral philosophy. The piece closes with a brief historical comparison of several leading ethical theories.

R. S. Peters's analysis of two major theories of justification—naturalism and intuitionism—also touches on the history of ethics. His comparison of *traditional* intuitionism, which emphasizes natural law and self-evident truths, with *modern* intuitionism, which postulates the existence of "non-natural" properties or qualities such as "goodness," is particularly instructive.

Peters also analyzes the strengths and weaknesses of the two positions under consideration.[1] He applauds the attempt on the part of naturalists to maintain a sense of objectivity by linking moral judgments to certain assumptions about human nature. But the very strength of naturalism reveals its main weakness, according to Peters, for in attempting to reduce values to facts, naturalism fails to do justice to the "autonomy" of ethics. Intuitionism, on the other hand, does preserve that autonomy, and therein lies its strength. Regardless of whether its focus is on "grasping relationships" (the traditional view) or on "inspecting qualities" (the modern view), however, intuitionism is inadequate, Peters holds, because it fails to provide a basis for translating moral perception into action.

Frankena's second selection in this section also deals with naturalism and intuitionism, along with emotivism and more recent (also more moderate) noncognitivist theories of justification. As part of his general discussion of meta-ethics, Frankena provides a detailed analysis of the strengths and weaknesses of these various theoretical positions. In addition, he offers a thoughtful criticism of relativism in ethics, presents his own theory of justification, and considers the fundamental ethical question, "Why should we be moral?"

Part One, then, provides a philosophical frame of reference for considering the value judgments that are made with regard to education in the remainder of the book.

NOTE

1. In a section of the chapter not included here, he also analyzes the emotive theory of justification. In this collection, emotivism is discussed in the second essay by Frankena in Part One.

1. Ethics

William K. Frankena

Ethics: (Gr. ta ethika, from ethos) *Ethics* (also referred to as moral philosophy) is that study or discipline which concerns itself with judgments of approval and disapproval, judgments as to the rightness or wrongness, goodness or badness, virtue or vice, desirability or wisdom of actions, dispositions, ends, objects, or states of affairs. There are two main directions which this study may take. It may concern itself with a psychological or sociological analysis and explanation of our ethical judgments, showing what our approvals and disapprovals consist in and why we approve or disapprove what we do. Or it may concern itself with establishing or recommending certain courses of action, ends, or ways of life as to be taken or pursued, either as right or as good or as virtuous or as wise, as over against others which are wrong, bad, vicious, or foolish. Here the interest is more in action than in approval, and more in the guidance of action than in its explanation, the purpose being to find or set up some ideal or standard of conduct or character, some good or end or *summum bonum*, some ethical criterion or first principle. In many philosophers these two approaches are combined. The first is dominant or nearly so in the ethics of Hume, Schopenhauer, the evolutionists, Westermarck, and of M. Schlick and other recent positivists, while the latter is dominant in the ethics of most other moralists.

Either sort of enquiry involves an investigation into the meaning of ethical statements, their truth and falsity, their objectivity and subjectivity, and the possibility of systematizing them under one or more first principles. In neither case is ethics concerned with our conduct or our ethical judgments simply as a matter of historical or anthropological record. It is, however, often said that the first kind of enquiry is not ethics but psychology. In both cases it may be said that the aim of ethics, as a part of philosophy, is theory not practice, cognition not action, even though it be added at once that its theory is for the sake of practice and its cognition a cognition of how to live. But some moralists who take the second approach do deny that ethics is a

Reprinted with permission of the publisher from the *Dictionary of Philosophy*, ed. Dagobert D. Runes (New York: Philosophical Library, Inc., 1960), pp. 98–100.

cognitive discipline or science, namely those who hold that ethical first principles are resolutions or preferences, not propositions which may be true or false, e.g., Nietzsche, Santayana, Russell.

Ethical judgments fall, roughly, into two classes, (a) judgments of value, i.e. judgments as to the goodness or badness, desirability or undesirability of certain objects, ends, experiences, dispositions, or states of affairs, e.g. "Knowledge is good," (b) judgments of obligation, i.e. judgments as to the obligatoriness, rightness or wrongness, wisdom or foolishness of various courses of action and kinds of conduct, judgments enjoining, recommending or condemning certain lines of conduct. Thus there are two parts of ethics, (1) the theory of value or axiology, which is concerned with judgments of value, extrinsic or intrinsic, moral or nonmoral, (2) the theory of obligation or deontology, which is concerned with judgments of obligation. In either of these parts of ethics one may take either of the above approaches—in the theory of value one may be interested either in analyzing and explaining (psychologically or sociologically) our various judgments of value or in establishing or recommending certain things as good or as ends, and in the theory of obligation one may be interested either in analyzing and explaining our various judgments of obligation or in setting forth certain courses of action as right, wise, etc.

Historically, philosophers have, in the main, taken the latter approach in both parts of ethics, and we may confine our remaining space to it. On this approach a theory of value is a theory as to what is to be pursued or sought, and a theory of obligation, a theory as to what is to be done. Now, of these two parts of ethics, philosophers have generally been concerned primarily with the latter, busying themselves with the former only secondarily, usually because it seemed to them that one must know what ends are good before one can know what acts are to be performed. They all offer both a theory of value and a theory of obligation, but it was not until the 19th and 20th centuries that value-theory became a separate discipline studied for its own sake—a development in which important roles were played by Kant, Lotze, Ritschl, certain European economists, Brentano, Meinong, von Ehrenfels, W. M. Urban, R. B. Perry, and others.

In the theory of value the first question concerns the meaning of value-terms and the status of goodness. As to meaning the main point is whether goodness is definable or not, and if so, how. As to status the main point is whether goodness is subjective or objective, relative or absolute. Various positions are possible. (a) Recent emotive meaning theories, e.g. that of A. J. Ayer, hold that "good" and other value-terms have only an emotive meaning. (b) Intuitionists and non-naturalists often hold that goodness is an indefinable intrinsic (and therefore objective or absolute) property, e.g., Plato, G. E. Moore, W. D. Ross, J. Laird, Meinong, N. Hartmann. (c) Metaphysical and naturalistic moralists usually hold that goodness can be defined in metaphysical or in psychological terms, generally interpreting "x is good" to mean that a certain attitude is taken toward x by some mind or group of minds. For some of them value is objective or absolute in the sense of having the same locus for everyone, e.g., Aristotle in his definition of the good as that at which all things aim, (*Ethics*, bk. I). For others the locus of value varies from individual to individual or from group to group, i.e. different things will be good for different individuals or groups, e.g., Hobbes, Westermarck, William James, R. B. Perry.

The second question in value-theory is the question "What things are good? What is good, what is the highest good, etc.?" On this question perhaps the main issue historically is between those who say that the good is pleasure, satisfaction, or some state of feeling, and those who say that the good is virtue, a state of will, or knowledge, a state of the intellect. Holding the good to be pleasure or satisfaction are some of the Sophists, the hedonists (the Cyrenaics, the Epicureans, Hobbes, Hume, Bentham, Mill, Sidgwick, Spencer, Schlick). Holding virtue or knowledge or both to be good or supremely good are Plato, Aristotle, the Stoics, the Neo-Platonists, Augustine, Aquinas, Spinoza, Kant, Hegel, G. E. Moore, H. Rashdall, J. Laird, W. D. Ross, N. Hartmann.

In the theory of obligation we find on the question of the meaning and status of right and wrong the same variety of views as obtain in the theory of value: "right," e.g., has only an emotive meaning (Ayer); or it denotes an intuited indefinable objective quality or relation of an act (Price, Reid, Clarke, Sidgwick, Ross, possibly Kant); or it stands for the attitude of some mind or group of minds towards an act (the Sophists, Hume, Westermarck). But it is also often defined as meaning that the act is conducive to the welfare of some individual or group—the agent himself, or his group, or society as a whole. Many of the teleological and utilitarian views mentioned below include such a definition.

On the question as to what acts are right or to be done ethical theories fall into two groups: (1) Axiological theories seek to determine what is right entirely by reference to the *goodness* or *value* of something, thus making the theory of obligation dependent on the theory of value. For a philosopher like Martineau it is the comparative goodness of its motive that determines which act is right. For a teleologist it is the comparative amount of good which it brings or probably will bring into being that determines which act is right—the egoistic teleologist holding that the right act is the act which is most conducive to the good of the agent (some Sophists, Epicurus, Hobbes), and the universalistic teleologist holding that the right act is the act which is most conducive to the good of the world as a whole (see *Utilitarianism*). (2) On deontological theories see *Deontological ethics* and *Intuitionism*.

Historically, one may say that, in general, Greek ethics was teleological, though there are deontological strains in Plato, Aristotle, and the Stoics. In

Christian moralists one finds both kinds of ethics, according as the emphasis is on the will of God as the source of duties (the ordinary view) or on the goodness of God as somehow the end of human life (Augustine and Aquinas), theology and revelation taking a central role in either case. In modern philosophical ethics, again, both kinds of ethics are present, with the opposition between them coming out into the open. Starting in the 17th and 18th centuries in Britain are both "intuitionism" (Cambridge Platonists, Clarke, Butler, Price, Reid, Whewell, McCosh, etc.) and utilitarianism (q.v.), with British ethics largely a matter of controversy between the two, a controversy in which the teleological side has lately been taken by Cambridge and the deontological side by Oxford. Again, in Germany, England, and elsewhere there have been, on the one hand, the formalistic deontologism of Kant and his followers, and, on the other, the axiological or teleological ethics of the Hegelian self-realizationists and the *Wertethik* of Scheler and N. Hartmann.

Ethical theories are also described as metaphysical, naturalistic, and non-naturalistic or intuitionistic. See *Intuitionism, Non-naturalistic ethics, Metaphysical ethics, Naturalistic ethics, Autonomy of ethics.*

Histories of Ethics: H. Sidgwick, *Outlines of the History of Ethics*, Rev. Ed. 1931. Gives titles of the classical works in ethics in passing. C. D. Broad, *Five Types of Ethical Theory*, 1930.

Elementary Texts: J. Dewey and J. H. Tufts, *Ethics*, Rev. Ed. 1932. W. M. Urban, *Fundamentals of Ethics*, 1930.

Treatises: H. Sidgwick, *Methods of Ethics*, 7th Ed. 1907. G. E. Moore, *Principia Ethica*, 1903. W. D. Ross, *Foundations of Ethics*, 1939. N. Hartmann, *Ethics*, 3 vol., trans. 1932. M. Schlick, *Problems of Ethics*, trans. 1939. R. B. Perry, *General Theory of Value*, 1926.—*W.K.F.*

2. Classical Theories of Justification

R. S. Peters

* * *

2. NATURALISM

An obvious suggestion, when confronted with questions about what ought to be done, is to assimilate them to the more straightforward questions about what is, was, or always will be the case. It has often been argued, for instance, that if man's nature could be clearly determined, then it would be clear how man ought to live. A whole succession of eminent philosophers have argued that man is essentially rational by nature. He differs from other animals in the development of reason as exhibited in his ability to impose plans and rules on his desires and in his capacity for abstract thought and the development of highly complicated symbolic systems; his behaviour can only be understood if account is taken of his rationality so defined. Therefore, it is argued, those activities are best which exhibit rationality in the highest degree. Mathematics is obviously preferable to bingo in that it satisfies more fully man's nature as a rational being.

An argument of a similar form is often advanced for the importance of personal relationships. It is argued that only men are capable of treating other members of their species in this way. Other animals may have innate or acquired social tendencies but they are incapable of appreciating their fellows, as distinct centres of consciousness and of encountering them as persons. Therefore, it is argued, man should live his life as much as possible on the plane of the personal rather than merely as an organism or as an individual occupying a purely functional position.

Reprinted with permission of the publisher from *Ethics and Education*, 2d ed. (London: George Allen & Unwin LTD, 1970), pp. 94–107.

Few civilized people—especially teachers—would wish to dispute the moral desirability of such policies for living. There may well be, too, some sort of connection between their desirability and man's rationality. But there are all sorts of difficulties about the form of argument in which such policies feature as a conclusion. There are difficulties, first of all about what is meant by 'nature' when any premiss is formulated about 'the nature of man'. In the arguments here cited 'nature' is interpreted as indicating some important respect in which man differs from other animals. An initial objection might be made on the score of the arbitrariness involved in the selection of man's capacity for reason or for forming personal relationships. No other living thing laughs like man or spends so long rearing its young; the possession of a prehensile thumb which enables man to use tools might be mentioned as being of more importance than his capacity for abstract thought. Genetically speaking this might well have developed from the ability to perform concrete operations made possible by the possession of such a thumb. Indeed Julian Huxley has indicated about twenty important ways in which man is different from other animals.[1] But *whatever* capacity is suggested as being crucial in determining man's difference from other animals the form of the argument remains the same. And it is difficult to see how the fact that man is different from animals in a certain respect constitutes, of itself, a reason for claiming that it is this respect that ought to be developed. Indeed Rousseau was willing to agree that man differs from animals in that he thinks, but went on to maintain that a thinking man is a depraved animal. The use of reason leads to an alienation of man from the rest of nature, which is undesirable. If contradictory conclusions can be 'inferred' from the same set of premisses there must be something wrong with the form of argument.

Others have argued that, because man is the same as other animals in respect of his aggression or need for a horde, it is these tendencies that ought to be developed. Philosophers have seldom advanced this as an argument. But this is surely because *for other reasons* they have been committed to a policy of rationality rather than brutishness. It cannot have been because a premiss which contains 'different from' rather than 'same as' as part of an empirical generalization is logically preferable.

This point can be generalized into an attack on simple naturalistic theories in ethics which try to base ethical recommendations on empirical generalizations. Naturalism is sometimes represented as the view that tries to infer moral judgments from statements of fact. But this way of characterizing naturalism is too loose; for the notion of what constitutes a fact is loose. Etymologically the word 'fact' intimates something that is palpably done. Facts can be contrasted with theories, with fiction, and with opinion.[2] So presumably the contrast arises between 'facts' and 'values' because what is valuable is thought to be a matter of opinion. But it is perfectly good English to say 'let us start from the fact that pain is evil' or 'it is a palpable fact that men ought to keep their promises'. Empirical observations, on the other hand, provide the most obvious hard core of what is *not* a matter of conjecture, fiction, or opinion. So it is easy to see how the notion of what is a matter of fact can be sucked up into the notion of what can be observed.

Goodness or desirability, however, are not things, relations, or qualities that can be observed to be present in man's activities, neither can statements about what is good or desirable be straightforwardly inferred from what can be observed. Empirical generalizations about man's nature, in so far as these involve judgments of comparison with animals, provide no basis for inference either. If it is argued that men *ought* to develop their reason because in this respect they are different from animals, this is only valid if the implicit principle is made explicit, that men ought to develop that capacity in which they differ from animals. Once this principle is made explicit, which makes man's reason or his possession of a prehensile thumb a relevant consideration, it is obvious that there is a problem about the justification of such a basic principle.

The invalidity of this form of argument does not derive simply from the interpretation of what is natural to man in terms of the comparison with animals. Any empirical generalization is subject to the same objection if it is put forward as a premiss unsupported by an ethical principle. Hobbes, for instance, argued that man ought to accept simple rules necessary for the preservation of peace because he was by nature afraid of death as well as zealous for power and advantage over other men. Mill tried to base an argument for the desirability of happiness on man's universal tendency to desire it. The logical difficulty about all such arguments is that answers to practical questions about what ought to be done are inferred from answers to theoretical questions about what is the case. As Hume put the matter at a time when men were beginning to get clearer about such forms of argument:

> In every system of morality which I have hitherto met with, I have always remarked that the author proceeds for some time in the ordinary way of reasoning, and establishes the being of God, or makes observations concerning human affairs; when, of a sudden I am surprised to find, that instead of the usual copulation of propositions, *is*, and *is not*, I meet with no proposition that is not connected with an *ought* or *ought not*. This change is imperceptible; but is however of the last consequence. For as this *ought* or *ought not* expresses some new relation or affirmation, it is necessary that it should be observed and explained; and at the same time that a reason should be given for what seems altogether inconceivable, how this new relation can be a deduction from others, which are entirely different from it.[3]

Hume is here asserting what is nowadays called the autonomy of ethics which is the claim that no moral judgment can be deduced from any set of premisses which does not itself contain a moral judgment or principle. Naturalism as an ethical theory ignores the requirement of a strict deductive argument that nothing should be drawn out by way of a conclusion that is not contained implicitly or explicitly in the premisses. If there is no practical principle, which enjoins some form of actions or expresses some preference, in the premisses, how can a judgment of this form ever emerge as a conclusion?

Consequentialist theories which attempt to base the desirability of courses of action on their consequences are open to the same objection unless an ethical principle is stated which makes some forms of consequences relevant. Supposing, for instance, it were argued that it was wrong to stick sharp instruments into other people because of the pain that was caused by so doing. This would only be a valid argument against this form of conduct on the assumption that pain is undesirable. Another person might argue that this was a most desirable form of conduct because it tended to make people bleed. Blood being red, this form of conduct tended to increase the amount of redness in a drab world. Hence its desirability. To most of us this would seem to be an argument which only a lunatic would advance. This is because the principle that pain ought to be minimized seems more perspicuous than the principle that redness ought to be maximized. But the argument, though bordering on lunacy, does at least bring out the point that an ethical principle is needed to pick out the relevance of consequences as well as the point that some principles for doing this seem more acceptable than others.

The naturalist might reply that the example brings out the point which he wishes to stress—that some consequences rather than others are relevant because of human nature. The fact is that human beings universally avoid pain whereas they do not universally seek redness. Hence the tendency to pick out consequences which involve pain rather than redness. There is something in this argument but not all that the naturalist might claim. For in general the existence of a human want is not a sufficient ground for maintaining the desirability of what is wanted. If Freud is right there might well be a universal desire amongst men to seduce their mothers and to kill their fathers. But the desirability of such forms of conduct would not depend on demonstrating the universality of the desires which might prompt it. Indeed, as Freud himself argued, one of the main functions of moral rules is to regulate such 'natural' desires. Nevertheless it would be very odd if there were *no* sort of connection between moral rules and human wants; for one of the distinguishing features of moral rules is that they guide conduct. If there were no connection between what they enjoined and some things which human beings tend to want the actual guiding function of practical discourse would be inexplicable.

The defect of naturalism is that it makes the connection between human nature and this guiding function of moral discourse too tight. It suggests that generalizations about human nature function as premisses from which rules of conduct can be inferred. Alternatively it is suggested (e.g. by Hobbes) that 'good' means 'that which a man desires'. This is a most implausible suggestion. For there is no logical contradiction in asserting that peace is good but that no one desires it. We can also, with perfect logical propriety, remark that people want things that are not worth wanting or that they want all the wrong things. Furthermore, when we bring up children we tell them that they ought to do things or that certain things are good when we know perfectly well that these are not things that they want to do or have.

If in reply it is asserted that 'good' means what a man wants on the whole and in the long run, when he carefully considers what is in his interest, there is then the problem of explicating these qualifications without smuggling in some norm by reference to which the variety of human wants can be assessed.[4]

It would be inconceivable, however, that words like 'good' and 'ought' could function as they do in a public language to guide people's behaviour if there were *no* connection between what was prescribed and what might be wanted. It would be odd, for instance, to say that something was good, such as making the world much redder, if it were inconceivable that anyone should want such a state of affairs.

The truth is surely that words such as 'good' and 'bad' are taught as words in a public language in connection with objects and states of affairs such as eating food and experiencing pain which are universal objects of desire or aversion. They could only have the guiding function which they do have if they were typically associated with such paradigm objects of desire and aversion. But once they have, as it were, got off the ground as words in a public language, they can be used to guide people towards things which they do not actually want but which are possible objects of desire. Indeed 'good' and 'ought' are used very frequently for educating people in respect of their wants, once they have grasped the commending function of such words.

Perhaps one of the strongest points of naturalism as an ethical theory is that it does do justice to what is usually called the 'objectivity' of moral judgments. By 'objectivity' is meant the assumptions that error is possible in moral matters and that whether or not a person is in error depends on facts independent of the opinions or attitudes of any particular person or group of persons. To claim objectivity is to deny that the adoption of moral values is merely a matter of personal taste or group allegiance. Words like 'ought', 'wrong', 'good', and 'bad' typically feature in a form of discourse which has not only the practical function of determining action but also the function of doing this by the production of reasons. To give reasons, if it is done seriously and with a determination to decide in terms of reasons, is to put a matter up for public discussion. It is tantamount to the admission that the decision must depend not on the authority or private whims of any individual but on the force and relevance of the reasons advanced. It is to assume, too, that truth and error are possible about the matter under discussion. For how could *discussion* have any point without such an assumption?

Naturalists are united in assuming that moral matters admit of such discussion and that the reasons adduced as backing to moral judgments must consist of empirical generalizations—usually to do with human nature or human wants. This is salutary in its insistence on the connection between moral discourse and the giving of reasons. But it falls short by ignoring the ethical principle or principles which are necessary to make such reasons relevant, and to bridge the gap which Hume detected between 'is' and 'ought'. In brief naturalism does justice to the objectivity of moral discourse; most forms of it do justice to its guiding function by connecting it with human nature or human wants; but it does not do justice to its autonomy.

3. INTUITIONISM

The strongest point of the second classical theory that has to be considered, which is usually called 'intuitionism', is that it does preserve the autonomy of ethics. In all its forms it is clear about two points, namely that terms like 'good' and 'ought' do not stand for observable qualities or relations and that moral judgments are not inferences from any form of empirical generalization. In this respect intuitionism is a great improvement on naturalism. Unfortunately, however, its critique of naturalism and the theory which is developed as an alternative, preserve the main feature of the theory which it criticizes. For it assimilates moral judgment to the providing of answers to a very special form of purely theoretical question. It supposes that in the end the goodness of an activity like artistic creation or the rightness of a principle such as that of justice or liberty is a matter of 'seeing' or grasping a quality or relation.

This process of 'seeing' has been interpreted in terms of two distinct sorts of models, both of which attempt to ground moral knowledge on some kind of indubitable and self-evident propositions. One type of theory assumes that terms like 'good' designate some sort of property that is grasped by a reflective mind. G. E. Moore,[5] for instance, a modern Platonist, held that 'good' refers to a simple non-natural unanalysable property. It is non-natural in the sense that it is not to be discovered by use of the senses or by the ordinary processes of introspection by which anyone might, for instance, become aware of thirst or pain. Nevertheless it depends for its existence on such observable qualities in that it is only to be discerned when some such observable qualities are also present. The goodness of artistic creation, for instance, is not something that can be observed by means of the senses; neither is it an inference that can be tested by observation. Nevertheless it is only discernible by the inner eye of intuition, when other observable qualities are present, in virtue of which it can be said, for instance, that a man is painting a picture. This theory uses the model of looking at a simple quality like yellow, which many have taken to be the sort of experience which is the ground of our certainty about the world and to issue in incorrigible statements like 'This is yellow.' It then postulates a special non-natural type of object or quality, a Platonic form, as the object of a special inward type of 'seeing'. Moral knowledge is thought to be based, in the end, on this intellectual type of grasping in the same sort of way as empirical knowledge is thought to be based on our sensory experience of simple qualities and relations. 'Intuition', being derived from the Latin 'intueor' which meant 'I gaze on', is the name given to this intellectual process of 'seeing' which is crucial for this type of theory.

When confronted with moral principles rather than with good activities, intuitionism often has employed another model of 'seeing' which is much more closely linked with the grasp of the self-evident which many once thought to lie at the root of mathematics. Mathematics was taken by Plato, and by later thinkers such as Descartes, as the paradigm of knowledge; so it was concluded that moral knowledge, if it is to be knowledge at all, must resemble mathematical knowledge in its structure. Mathematical knowledge, especially geometry, was thought to be based on a clear and distinct apprehension of basic Forms or simple natures between which relations could be grasped intuitively. From this self-evident foundation of axioms demonstrations could be made which issued in theorems. Descartes held that certain knowledge was possible in science and morals as well as in mathematics provided that this type of logical structure of propositions could be articulated.

This view of moral knowledge was later held by the post-Renaissance theorists of natural law. In the hands of John Locke it formed the epistemological basis for the conviction that there are certain inviolable rights of man—to life, liberty, and estate—a conviction which later provided a rationale for revolutionary doctrines as well as for the American Declaration of Independence. In more recent times moral philosophers such as Sir David Ross[6] have held that all moral duties are founded upon a limited number of basic prima facie obligations, such as that promises ought to be kept, which are known intuitively.

This form of intuitionism, then, which had its source in the importance ascribed by Plato to mathematics as a form of knowledge, has had a long and influential history in the development of ethics. There are, nevertheless, many basic objections both to this type of intuitionism and to the type stated so precisely by G. E. Moore, which employed the model of inspecting qualities rather than that of grasping relationships, to characterize the foundations of moral knowledge.

To start with, the view that mathematics itself must be based on intuitively grasped axioms is not now widely accepted. Furthermore, since the attacks by Hume and Kant on Cartesian rationalism, the fact that mathematical systems sometimes fit the world presents itself as a problem, the rationalistic assumption having been abandoned that mathematical thinking must somehow mirror the world, or at least the real structure of the world beneath the appearances, which was what Plato assumed in his theory of Forms and Descartes in his theory of simple natures. Whether or not a postulate about the world is true depends, in the end, on whether consequences deduced from it can be confirmed by observation, not on the self-evidence of the postulate itself.

The intuitionist view usually goes hand in hand with some notion of self-evidence which is alleged to characterize the grasp of both mathematical and moral axioms. 'Self-evidence' is a term which combines both logical and psychological attributes. Psychologically speaking some sort of inner flash is alleged to occur when such axioms are grasped; logically speaking the 'evidence' is thought of as being internal to what is affirmed. The logical requirement is most easily exhibited in statements when the truth of what is affirmed derives from the rules governing the terms which are combined. Examples would be 'a square contains some right angles' or 'every effect has a cause'. The truth of statements about the world, however, does not depend in the end upon such conventions for the use of terms. The definition of the term 'father' does not reveal the truth of the Freudian assumption that fathers are hated by their sons in the way which it reveals the truth of the statement that if a man is a father he must have either a son or a daughter. Of course statements about the world can be so well confirmed that an internal relation of this sort can grow up. For instance 'gold is yellow' has been found so often to be true that we would not be inclined to call anything a piece of gold which did not reveal this colour under standard conditions. But this connection had to be established by observation; it was not created by convention like some of the truths of mathematics. People may also come to experience a feeling of conviction about such statements. But this feeling depends upon the truth of the statement being established in some other way; it is an accompaniment of a statement being true, not a criterion of its truth. It may be useful to have such feelings of conviction; for they curtail the area of doubt and leave the mind free to speculate about a limited realm of problems. But an aid to practical living should not be exalted into a criterion of truth.

Some moral judgments such as 'murder is wrong' may come to have such a feeling attached to them. Many such judgments, too, are almost true by convention in the sense that 'murder' is almost equated with 'wrongful killing' in the public consciousness, whatever its legal definition may be. But this only puts the problem back as to what makes the sort of killing, picked out by the term 'murder', wrong. Definitions cannot of themselves determine what is right or wrong. Still less can the strength of people's feelings of certainty or 'self-evidence'. Many people feel just as strongly about abortion or about homosexuality between consenting males. Their feelings are regarded as highly arbitrary by others who make different judgments about these matters.

The basic objection, therefore, to intuitionism as an ethical theory is that the probing for fundamental principles is encouraged to stop too soon and at too *arbitrary* a point. If it is put forward, as by Sir David Ross, that certain basic obligations are self-evident, there are many who may doubt this. A good example is the case of punishment where many have hotly challenged the intuitive conviction that pain ought to be inflicted on those who commit breaches of rules. And even if they agree, for instance, that promises ought in general to be kept they may deny that this is 'self-evident'; for, they will argue, there are very good reasons for keeping promises. Can it be seriously maintained that there are no further reasons for this important social practice? If obligations like this are held to be self-evident, what is to be said to people who insist that black men ought to be treated differently just because they are black, that one ought never to permit more redness in the world than is necessary, or that gambling is self-evidently wrong? They can say that they just know these things intuitively in the same sort of way as intuitionist philosophers claim that they know intuitively that they ought not to tell lies, break their promises, and be unfair. The attempt to justify principles stops at an arbitrary point. Where arbitrariness reigns, *de gustibus non est disputandum*. This is a very unsatisfactory basis for morality which intuitionists have always insisted is an impersonal and objective matter.

The other type of intuitionism, which likens being aware of goodness to 'seeing' in a more literal sense, is no better off in respect of the accusation of arbitrariness either. In ordinary cases of 'seeing' there are criteria for making the distinction between something being 'really' there or as it is described and something only 'appearing' to be so. Reference to standard conditions, perspective, etc., can be made. But what are standard conditions for 'seeing' the non-natural quality of goodness, which permit a similar escape from arbitrariness? Furthermore in ordinary cases of seeing there are established tests for determining whether there is something wrong with the observer if he fails to see what is there or sees things askew or is subject to hallucinations. What similar tests are there in the moral case? Can a man be assessed as morally blind in the same sort of way as he can be assessed as colour-blind?[7]

Both types of intuitionism suffer from a further fundamental defect which has been already hinted at and which was first stressed by David Hume. The stress on 'seeing', whether modelled on seeing qualities or grasping relations, makes morals a theoretical matter. If a man sees that a table is square or grasps a mathematical truth, there is no implication of any sort about anything being done. He may remark 'How interesting!' after noting what there is to note or grasping what there is to grasp. But whether or not anything is to be done is a further question. Now the palpable thing about the use of moral concepts such as 'good', 'wrong', and 'ought', is that there is some kind of close connection with something being done; they are used to guide people's choices. The precise form of this connection is difficult to determine; but that there is such a connection is one of the basic features of moral language. The intuitionists make the connection between seeing that something is good or grasping that something ought to be and doing something about it a purely contingent one.

Plato was one of the first to suggest an intuitive basis for moral knowledge. But he did not think of the connection with action as a contingent one. For he inherited from Socrates the conviction that virtue is knowledge and that no man wittingly does what he knows to be evil. He therefore believed that knowledge of what is good is accompanied by a passionate desire that it should be realized. This Socratic doctrine seems to err in the other direction; for it makes the connection between moral knowledge and action too close. It does not seem to be a logical contradiction for a man to say that he knows what is good but that he does not want or intend to do it. Yet Socrates maintained that such a man could not really know what is good.

The Socratic view, however, seems more defensible than that of the later intuitionists who assimilated moral knowledge to perception or to intellectual assent. For it does make the connection between moral knowledge and action, albeit in too tight a manner. The true view is surely that the *general function* of moral language is to guide action or to get people to do things. To say that something is good is to intimate that there are reasons for doing or promoting it. These must be reasons which could induce human beings of some sort to act. Otherwise moral language could never have the function which it does have in getting people to do things. But the reasons intimated in a particular case may not, psychologically speaking, be strong enough to get the person who is using or addressed in the public language to act in the way prescribed. He may say or agree that peace is good—meaning that there are very good reasons for having peace. But he may know well enough that he personally does not want it strongly enough to pursue it himself. Nevertheless it could not sensibly be said that it was good if there was nothing about it which would count, for some human being or other, as a reason for trying to promote it.

Though intuitionism, like naturalism, is inadequate as an ethical theory it nevertheless, like naturalism, intimates in a distorted way some of the important features of moral knowledge and discourse. It will be useful, therefore, at this point to make explicit what these features are. The point was made, in the foregoing criticism of intuitionism, that words like 'good' and 'ought' occur in a form of discourse that has the practical function of guiding people's action. This is not the only way in which people's behaviour is regulated by language. Another common form of practical discourse is the use of commands. What distinguishes moral language, in which terms like 'good' and 'ought' occur, is the suggestion that there are reasons for doing what is prescribed. 'Shut the door' and 'You ought to shut the door' both have a practical function. But 'ought' implies reasons for shutting the door in a way in which the use of imperatives does not. Nothing, of course, depends on the use of the particular words 'good' or 'ought' to perform this function. Once language has become sufficiently differentiated so that getting people to do things by giving them orders has become distinguished from getting them to do things by giving them reasons, any words could have done. So this is not a verbal argument in the sense that anything of substance is inferred from the use of particular words. Rather it is an argument which goes behind words and attempts to formulate in a general way what is presupposed by the differentiation of forms of discourse. What is presupposed, it is claimed, is the development of the concept of getting people to do things by giving them reasons, which is a very specific form of regulation associated with a specific family of words. 'Good', 'bad', 'ought', 'ought not', 'right', 'wrong' are the most general words in our vocabulary for performing this function. 'Intuitionism' as an ethical theory marks the point very strongly that reasoning in morals, or practical reason, is very different in respect of its justification from reasoning about what is, was or will be the case, for which the descriptive and explanatory discourse of science and history has been developed. But it does this in a very misleading way; for it preserves the model appropriate for answering theoretical questions by basing moral knowledge on 'seeing' in one of the two forms outlined.

This model caricatures another of the main features of practical discourse which is often referred to as its autonomy—the fact that statements about what is good or what ought to be done cannot be reduced to or inferred from empirical observations or generalizations. The postulation of 'non-natural' qualities and relations and of an inner eye for inspecting or grasping them make this point in a dramatic but misleading way. For by suggesting that the grounds of moral knowledge are grasped by a kind of gazing it ignores the practical function of moral discourse, its very close connection with action. In this respect, therefore, intuitionism employs a misleading model to make an important point about the analysis of moral terms and the justification of moral convictions.

In another respect, too, intuitionism makes an important point about moral discourse, though again in a misleading way. To claim that moral discourse is that form of practical discourse whose function is to regulate behaviour by the giving of reasons is, *ipso facto*, to claim some sort of *objectivity* for it. The claim that there are reasons for what is prescribed distinguishes moral language from that in which private likes and whims are canvassed. 'I like bull-fighting' conveys something very different from 'I approve of bull-fighting' just as saying that gardening is good is very different from saying that it is nice. Intuitionism attempts to convey this claim of objectivity, of interpersonal standards, by postulating the paradigm case of objectivity where some quality or relation is out there for all to behold. As has been shown it cannot sustain the claim to objectivity in this particular form. For 'non-natural' qualities and relations seen by an inner eye present problems of standard conditions and normal observers. But at any rate intuitionism does hang grimly on to this claim of objectivity. Rightly so; for this is implicit in the very notion of there being a reason for doing something.

Again, when this notion of the backing of reasons is pressed, both the strength and weakness of intuitionism are again revealed. Intuitionists, especially Plato, have always stressed that the ultimate experience of 'seeing' is only granted to reflective people who think detachedly and clearly about situations and activities. The 'intuition' provides, as it were, the basis beyond which further reasoning is impossible. Indeed the general function of saying that something is known 'intuitively' is to suggest that what is claimed is true, that one is sure that it is true, but to withdraw the usual claim that goes with 'know', that the grounds for saying it is true are patent. In this way intuitionism preserves the link with reason, but gives up too soon in the search for a foundation to morals in principles of a non-arbitrary sort. What these principles are in the sphere of practical reason and how they can be shown to be nonarbitrary must be left for later consideration.

* * *

NOTES

1. Huxley, J., *The Uniqueness of Man* (London, Chatto and Windus, 1941).
2. For a most illuminating discussion of the category of 'fact' see Hamlyn, D., 'The Correspondence Theory of Truth' in *Philosophical Quarterly*, July 1962.

3. Hume, D., *Treatise on Human Nature*, Book III, Part I, Section 2.
4. See Griffiths, A. P., and Peters, R. S., 'The Autonomy of Prudence', *Mind*, April 1962.
5. Moore, G. E., *Principia Ethica* (Cambridge University Press, 1903).
6. Ross, Sir David, *The Right and the Good* (Oxford, The Clarendon Press, 1930).
7. For further developments of this type of objection see Nowell-Smith, P. H., *Ethics* (Harmondsworth, Penguin Books, 1954), Ch. 1.

3. Meaning and Justification

William K. Frankena

META-ETHICS AND ITS QUESTIONS

Thus far, except for Chapter 1, we have been engaged in normative ethics, although we have also included a good bit of analysis and conceptual clarification, as well as some psychology. In other words, we have been endeavoring to arrive at acceptable principles of obligation and general judgments of value in the light of which to determine what is morally right, wrong, or obligatory, and what or who is morally good, bad, or responsible. As we saw in Chapter 1, however, ethics also includes another kind of inquiry called meta-ethics. Meta-ethics does not propound any moral principles or goals for action, except possibly by implication; as such it consists entirely of philosophical analysis. In fact, recent moral philosophy has concerned itself very largely with meta-ethical analysis; it has been primarily interested in clarification and understanding rather than in normative ethics, though it has included some discussion of punishment, civil disobedience, war, etc., and much debate about utilitarianism. For all that, what it has been doing is most important, since any reflective person should have some understanding of the meaning and justification of his ethical judgments, especially in this age when our general thinking about principles and values is said to be in a state of crisis. In any case, we ourselves must see what sort of justification, if any, can be claimed for the normative positions we have taken.

As usually conceived, meta-ethics asks the following questions. (1) What is the meaning or definition of ethical terms or concepts like "right," "wrong," "good," "bad"? What is the nature, meaning, or function of judgments in

William K. Frankena, *Ethics*, 2d ed. © 1973, pp. 95–116. Reprinted by permission of Prentice-Hall, Inc., Englewood Cliffs, New Jersey.

which these and similar terms or concepts occur? What are the rules for the use of such terms and sentences? (2) How are moral uses of such terms to be distinguished from nonmoral ones, moral judgments from other normative ones? What is the meaning of "moral" as contrasted with "nonmoral"? (3) What is the analysis or meaning of related terms or concepts like "action," "conscience," "free will," "intention," "promising," "excusing," "motive," "responsibility," "reason," "voluntary"? (4) Can ethical and value judgments be proved, justified, or shown valid? If so, how and in what sense? Or, what is the logic of moral reasoning and of reasoning about value? Of these (1) and (4) are the more standard problems of meta-ethics; but (2) and (3) have been receiving much attention lately. We have touched a little on all of them, but will now concentrate on (1) and (4).

Of these two problems, it is (4) that is primary. What we mainly want to know is whether the moral and value judgments we accept are justified or not; and if so, on what grounds. Question (1) is not in itself important in the same way. Apart from conceptual understanding—which is important to the pure philosopher—we need to be concerned about the meaning or nature of ethical and value judgments only if this helps us to understand whether and how they may be justified, only if it helps us to know which of them are acceptable or valid. I shall therefore state and discuss the main answers to question (1) if and when they are relevant to the discussion of question (4). It is not easy to classify all of the different theories of the meaning of ethical and value terms and judgments, but they seem to fall under three general types: *definist* theories, *intuitionism* or *non-naturalism*, and *noncognitive* or *nondescriptivist* theories. I shall explain them as they become relevant.

For the purposes of such discussions as these, moral judgments and nonmoral normative judgments are usually lumped together. This is a risky procedure, for it may be that rather different accounts must be given of the meaning and justification of the two kinds of judgments. Nevertheless, for convenience, we too shall adopt this procedure in our review of the various meta-ethical theories, and use the expression "ethical judgments" to cover all relevant normative and value judgments, not just moral ones.

THEORIES OF JUSTIFICATION

One way of putting question (4) is to ask whether our basic ethical judgments can be justified in any objective way similar to those in which our factual judgments can be justified. It is, therefore, by a natural impulse that many philosophers have sought to show that certain ethical judgments are actually rooted in fact or, as it used to be put, in "the nature of things" as this is revealed either by empirical inquiry, by metaphysical construction, or by divine revelation. How else, they ask, could one possibly hope to justify them as against rival judgments? If our chosen ethical judgments are not based on fact, on the natures and relations of things, then they must be arbitrary and capricious or at best conventional and relative. One who follows this line of thought, however, seems to be committed to claiming that ethical judgments can be derived logically from factual ones, empirical or nonempirical. Opponents have therefore countered by contending that this cannot be done, since one cannot get an Ought out of an Is or a Value out of a Fact.

Now, we do sometimes seem to justify an ethical judgment by an appeal to fact. Thus, we say that a certain act is wrong because it injures someone, or that a certain painting is good because it has symmetry. However, it becomes clear on a moment's thought that our conclusion does not rest on our factual premise alone. In the first case, we are tacitly assuming that injurious acts are wrong, which is a moral principle; and in the second, that paintings with symmetry are good, which is a value judgment. In such cases, then, we are not justifying our original ethical judgment by reference to fact alone but also by reference to a *more basic* ethical premise. The question is whether our *most basic* ethical or value premises can be derived logically from factual ones alone.

This would mean that conclusions with terms like "ought" and "good" in them can be logically inferred from premises, none of which contain these terms; this simply cannot be done by the rules of ordinary inductive or deductive logic. To try to do so is essentially to argue that A is B, $\therefore$ A is C, without introducing any premise connecting B and C. In this sense, those who insist that we cannot go from Is to Ought or from Fact to Value are correct. Such an inference is logically invalid unless there is a special third logic permitting us to do so. It has, in fact, been suggested by some recent writers that there is such a special logic sanctioning certain direct inferences from factual premises to conclusions about what is right or good, that is, an ethical logic with "rules of inference" like "If X is injurious, then X is wrong." But the theory and the rules of such a logic have not yet been satisfactorily worked out, and until they are we can hardly take this possibility seriously. In any case, it is hard to see how such a "rule of inference" differs in substance from the "premise" that injurious acts are wrong, or how its justification will be different.

DEFINIST THEORIES, NATURALISTIC AND METAPHYSICAL

There is, however, one possibility that must be taken seriously. This is the definist view that Ought can be defined in terms of Is, and Value in terms of Fact. For if such definitions are acceptable, then, by virtue of them, one can go logically from Is to Ought or from Fact to Value. For example, if "We ought to do . . . " means "We are required by society to do . . ., " then from "Society requires that we keep promises," it follows that we ought to keep promises. It will not do to reply, as some have, that no such definitions are possible since we cannot get an Ought out of an Is, for that is to beg the question. We must, therefore, take a closer look at definist theories.

According to such theories ethical terms can be defined in terms of nonethical ones, and ethical sentences can be translated into nonethical ones of a factual kind. For example, R. B. Perry proposes such definitions as these:

"good" means "being an object of favorable interest (desire),"
"right" means "being conducive to harmonious happiness."[1]

For him, then, to say that X is good is simply another way of saying that it is an object of desire, and to say that Y is right is just another way of saying that it is conducive to harmonious happiness. A theologian might claim that "right" means "commanded by God"; according to him, then, saying that Y is right is merely a shorter way of saying that it is commanded by God. On all such views, ethical judgments are disguised assertions of fact of some kind. Those who say, as Perry does, that they are disguised assertions of empirical fact are called *ethical naturalists*, and those who regard them as disguised assertions of metaphysical or theological facts are called *metaphysical moralists*.[2] Many different theories of both kinds are possible, depending on the definitions proposed. In each case, moreover, the definition presented may be advanced as a *reportive* one, simply explicating what we ordinarily mean by the term being defined, or as a *reforming proposal* about what it should be used to mean. Perry's definitions are offered as reforming proposals, since he thinks our ordinary use of "good" and "right" is confused and vague. F. C. Sharp, on the other hand, offers the following as reportive definitions:

"good" means "desired upon reflection,"
"right" means "desired when looked at from an impersonal point of view."[3]

In offering definitions or translations of ethical terms and judgments, a definist also tells us how such judgments are to be justified. For example, when Perry tells us that "good" means "being an object of desire," he also tells us that we can test empirically whether X is good simply by determining whether it is desired or not. In general, on a naturalistic theory, ethical judgments can be justified by empirical inquiry just as ordinary and scientific factual statements can; and on any metaphysical theory, they can be justified by whatever methods one can use to justify metaphysical or theological propositions. Either way they are rooted in the nature of things.

Opponents of such theories, following G. E. Moore, accuse them of committing "the naturalistic fallacy," since they identify an ethical judgment with a factual one. To call this a fallacy, however, without first showing that it is a mistake, as is sometimes done, is simply to beg the question. The critics also claim, therefore, that all proposed definitions of "good" and "right" in nonethical terms can be shown to be mistaken by a very simple argument, sometimes referred to as the "open question" argument. Suppose that a definist holds that "good" or "right" means "having the property P," for example, "being desired" or "being conducive to the greatest general happiness." Then, the argument is that we may agree that something has P, and yet ask significantly, "But is it good?" or "Is it right?" That is, we can sensibly say, "This has P, but is it good (or right)?" But if the proposed definition were correct, then we could not say this sensibly for it would be equivalent to saying, "This has P, but has it P?" which would be silly. Likewise, one can say, "This has P but it is not good (or right)," without contradicting oneself, which could not be the case if the definition were correct. Therefore the definition cannot be correct.

To this argument stated in such a simple form, as it almost always is, a definist may make several replies. (1) He may argue that the meaning of words like "good" and "right" in ordinary use is very unclear, so that when a clarifying definition of one of them is offered, it is almost certain not to retain all of what we vaguely associate with the term. Thus, the substitute cannot seem to be entirely the same as the original, and yet may turn out to be an acceptable definition. (2) He may point out that the term being defined may have a number of different uses, as we saw in the case of "good." Then P may be correct as an account of one of its uses, even though one can still say, "This has P, but is it good?" For one can agree, say, that X is good intrinsically, and still ask sensibly if it is good extrinsically, morally, or on the whole. (3) What we mean by some of our terms is often very hard to formulate, as Socrates and his interlocutors found. This means that one who doubts a certain formulation can always use the open question kind of an argument, but it does not mean that no definition can possibly be correct. (4) A definist like Perry may reply that the open question argument does show that the proposed definitions are not accurate accounts of what we mean by "good" and "right" in ordinary discourse, but that it still may be desirable to adopt them, all things considered. (5) A definist like Sharp, who thinks that his definitions do express what we actually mean, might even say that we cannot really ask significantly, "Is what we desire on reflection good?" or "Is what we approve when we take an impersonal point of view right?" His definitions are just plausible enough to give such a reply considerable force. In any case, however, although his critics may still be right, they will merely be begging the question if they rest their case on the open question argument.

The open question argument as usually stated, then, is insufficient to refute all definist theories. Its users almost never, in fact, make any serious effort to see what definists might say in reply or to consider their definitions seriously, as some of them certainly deserve to be. We cannot ourselves, however, try to consider separately all of the more plausible definitions which have been proposed. Even after studying them I find myself doubting that any pure definist theory, whether naturalistic or metaphysical, can be regarded as adequate as an account of what we do mean. For such a theory holds that an ethical judgment simply is an assertion of fact—that ethical terms constitute merely an alternative vocabulary for reporting facts. It may be that they should be reinterpreted so that this is the case. In actual usage, however, this seems clearly not to be so. When we are making merely factual assertions we are not thereby taking any pro or con attitude toward what we are talking about; we are not recommending it, prescribing it, or anything of the sort. But when we make an ethical judgment we are not neutral in this way; it would seem paradoxical if one were to say "X is good" or "Y is right" but be absolutely indifferent to its being sought or done by himself or anyone else. If he were indifferent in this way, we would take him to mean that it is generally regarded as good or right, but that he did not so regard it himself. We may be making or implying factual assertions in some of our ethical judgments—when we say, "He was a good man," we do seem to imply that he was honest, kind, etc.—but this is not all that we are doing.

It might be replied, by Perry for example, that we ought to redefine our ethical terms so that they merely constitute another vocabulary for reporting certain empirical or metaphysical facts (perhaps on the ground that then our ethical judgments could be justified on the basis of science or metaphysics). Then we would have to consider whether we really need such an alternative way of reporting those facts, and whether we can get along without a special vocabulary to do what we have been using our ethical terms to do—which at least includes expressing pro or con attitudes, recommending, prescribing, evaluating, and so on.

It seems doubtful, then, that we can be satisfied with any pure definist theory of the meaning of moral and other value judgments. It also seems to me that such theories do not suffice to solve the problem of justification. If we accept a certain definition of "good" or "right," then, as we saw, we will know just how to justify judgments about what is good or right. But this means that the whole burden rests on the definition, and we may still ask how the definition is justified or why we should accept it. As far as I can see, when Perry tries to persuade us to accept his definition of "right," he is in effect persuading us to accept, as a basis for action, the ethical principle that what is conducive to harmonious happiness is right. He cannot establish his definition first and then show us that this principle is valid because it is true by definition. He cannot establish his definition unless he can convince us of the principle.

This may seem obvious, since Perry's definition is meant as a recommendation. But a definist who regards his definition as reportive, and not reforming, would presumably rejoin by saying that his definition is justified simply by the fact that it expresses what we ordinarily mean, just as dictionary definitions are justified. It has been claimed that the notion of obligation as we know it was not present in Greek times and is due to the Judeo-Christian theology. It might be held, then, that "ought," as it is actually used in our moral discourse, means "commanded by God," and many people would accept this as an account of what they mean. If we ask such a reportive theological definist why we ought to do what God commands, he will probably answer, if he understands us to be asking for a justification and not for motivation, that we ought to do this because "ought" simply means "commanded by God." But this, if true, would only show that his ethical principle had become enshrined in our moral discourse; it would not show why we should continue to give adherence to his principle, and this is the question. In other words, to advocate the adoption of or continued adherence to a definition of an ethical term seems to be tantamount to trying to justify the corresponding moral principle. Appealing to a definition in support of a principle is not a solution to the problem of justification, for the definition needs to be justified, and justifying it involves the same problems that justifying a principle does.

If this is true, then our basic ethical norms and values cannot be justified by grounding them in the nature of things in any strictly logical sense. For this can be done logically only if "right," "good," and "ought" can be defined in nonethical terms. Such definitions, however, turn out to be disguised ethical principles that cannot themselves be deduced logically from the nature of things. It follows that ethics does not depend *logically* on facts about man and the world, empirical or nonempirical, scientific or theological.

It still may be that there is some *non-logical* sense in which our basic norms and value judgments can be justified by appeal to the nature of things. We have already seen that ethical egoists seek to justify their theory of obligation by arguing that human nature is so constituted that each of us always pursues only his own good, and that Mill and other hedonists try to justify their theory of value by showing that human nature is so constituted as to desire nothing but pleasure or the means to pleasure. Neither the egoists nor the hedonists claim that their argument affords a strict logical proof. I have also indicated that such arguments nevertheless have a very considerable force, provided their premises are correct. But we saw reason to question the premises of the psychological arguments for egoism and hedonism, and hence must take them as inadequate. In any case, however, it is doubtful that one could find any similar "proofs" of principles like beneficence, justice, or utility.

Many people hold that morality depends on religion or theology—that ethical principles can be justified by appeal to theological premises and only by appeal to such premises. To those who hold this we must reply, in view of our argument, that this dependence cannot be a *logical* one. They may, of course, still maintain that morality is dependent on religion in some psychological way, for example, that no adequate motivation to be moral is possible without religion. This, I think, is true, if at all, only in a very qualified sense; however, even if religious beliefs and experiences are necessary for *motivation*, it does not follow that the *justification* of moral principles depends on such beliefs and experiences. Theologians may also contend that the law of love or beneficence can be rationally justified on theological grounds, even if it cannot rest on such grounds logically. They may argue, for instance, that if one fully believes or unquestionably experiences that God is love, then one must, if he is rational, conclude that he, too, should love. They may say that, although this conclusion does not follow logically, it would be unreasonable for one to draw any other or to refrain from drawing it. In this belief they may well be right; for all that I have said, I am inclined to think they are right. However, it does not follow that the principle of beneficence (let alone that of equality) *depends* on religion for its justification even in this non-logical sense. It may be that it can also be justified in some other way.

INTUITIONISM

We must, then, give up the notion that our basic principles and values can be justified by being shown to rest *logically* on true propositions about man and the world. We may also have to admit or insist that they cannot be justified satisfactorily by any such psychological arguments as are used by egoists and hedonists. But now another familiar answer to the question of justification presents itself—the view that our basic principles and value judgments are intuitive or self-evident and thus do not need to be justified by any kind of argument, logical or psychological, since they are self-justifying or, in Descartes's words, "clearly and distinctly true." This view was very strong until recently, and is held by many of the writers we have mentioned: Butler, Sidgwick, Rashdall, Moore, Prichard, Ross, Carritt, Hartmann, Ewing, and possibly even by Plato. It is sometimes called *intuitionism*, sometimes *non-naturalism*.

Intuitionism involves and depends on a certain theory about the meaning or nature of ethical judgments. Definist theories imply that ethical terms stand for properties of things, like being desired or being conducive to harmonious happiness, and that ethical judgments are simply statements ascribing these properties to things. Intuitionists agree to this, but deny that the properties referred to by words like "good" and "ought" are definable in nonethical terms. In fact, they insist that some of these properties are indefinable or simple and unanalyzable, as yellowness and pleasantness are. Sidgwick holds that "ought" stands for such a property, Moore that "good" does, and Ross that both do. These properties are not, therefore, unintelligible or unknown, anymore than pleasantness and yellowness are. But they are not natural or empirical properties as are pleasantness and yellowness. They are of a very different kind, being non-natural or nonempirical and, so to speak, normative rather than factual—different in kind from all the properties dreamed of in the philosophies of the definists. According to this view, as for the definists, ethical judgments are true or false; but they are not factual and cannot be justified by empirical observation or metaphysical reasoning. The basic ones, particular or general, are self-evident and can only be known by intuition; this follows, it is maintained, from the fact that the properties involved are simple and non-natural.

On this view, ethical judgments may be and are said to be rooted in the natures and relations of things, but not in the sense that they can be derived from propositions about man and the world, as the views previously discussed hold. They are based on the natures and relations of things in the sense that it is self-evident that a thing of a certain nature is good, for example, that what is pleasant or harmonious is good in itself; or that a being of a certain nature ought to treat another being of a certain nature in a certain way, for example, that one man ought to be just, kind, and truthful toward another man.

There are a number of reasons why intuitionism, for almost two centuries the standard view among moral philosophers, now finds few supporters. First of all, it raises several ontological and epistemological questions. An intuitionist must believe in simple indefinable properties, properties that are of a peculiar non-natural or normative sort, a priori or nonempirical concepts, intuition, and self-evident or synthetic necessary propositions. All of these beliefs are hard to defend. Do our ethical terms point to distinct and indefinable properties? It is not easy to be sure, and many philosophers cannot find such properties in their experience. It is also very difficult to understand what a non-natural property is like, and intuitionists have not been very satisfying on this point. Moreover, it is very difficult to defend the belief in a priori concepts and self-evident truths in ethics, now that mathematicians have generally given up the belief that there are such concepts and truths in their field.

Intuitionism is also not easy to square with prevailing theories in psychology and anthropology, even if we do not regard relativism as proved by them, a point we will take up later. An enriched view of the meanings of meaning and of the functions and uses of language likewise casts doubt on the view that ethical judgments are primarily property-ascribing assertions, as intuitionists, like definists, believe.

Intuitionism may still be true in spite of such considerations. But there are two arguments against it that many have regarded as decisive. Both are used by noncognitivists or nondescriptivists and, interestingly enough, the first is similar to the open question argument used against definists by intuitionists themselves. Let us suppose, it is said, that there are such brave non-natural and indefinable properties as the intuitionists talk about. Let us also suppose that act A has one of these properties, P. Then one can admit that A has P and still sensibly ask, "But why should I do A?" One could not do this if "I should do A" means "A has P"; hence it does not mean "A has P" as intuitionists think.

I do not find this argument convincing. "Why should I do A?" is an ambiguous question. One who asks it may be asking, "What motives are there for my doing A?" or he may be asking, "Am I really morally obligated to do A?" That is, he may be asking for *motivation* or he may be asking for *justification*. Now, of course, one can admit that A has P and still sensibly ask, "What motives are there for my doing that which has P?" But this, an intuitionist may say, is irrelevant, since he is proposing a theory of justification and not a theory of motivation, although he is also ready to provide a theory of motivation at the proper time. Therefore, the question is whether one can admit that A has P and still ask sensibly, "Ought I really to do A?" Here we must remember that the intuitionist holds that "I ought to do A" *means* "A has P" or, in other words, that P *is* the property of obligatoriness. Hence, he can answer the argument in its relevant form by saying that *if* "I ought to do A" does mean that "A has P," then one cannot sensibly say, "A has P but ought I to do it?" His critic may still insist that he can sensibly say this, but not if he first admits that "I ought to do A" means "A has P." For him simply to assert that it does not mean "A has P" is to beg the question; however, his argument does not prove his conclusion, but assumes it. If there is a property of obligatoriness, as the intuitionist holds, then one cannot sensibly admit that A has this property and ask, "But is it obligatory?"

The second argument, which comes from Hume, is used against many kinds of definism as well as intuitionism, and has to do with motivation rather than justification. It begins with an insistence that ethical judgments are in themselves motivating or "practical" in the sense that, if one accepts such a judgment, he must have some motivation for acting according to it. It then contends that, if an ethical judgment merely ascribes a property, P, to something, then, whether P is natural or non-natural, one can accept the judgment and still have no motivation to act one way rather than another.

Intuitionists (and definists) also have a possible answer to this argument. They can maintain that we are so constituted that, if we recognize X to be right or good (i.e., that X has P), this will generate a pro attitude toward X in us, either by itself or by way of an innate desire for what has P. One may, of course, question their psychological claims, but one must at least give good reasons for thinking these are false before one takes this argument as final.

On the whole, however, intuitionism strikes me as implausible even if it has not been disproved. As was indicated earlier, ethical judgments do not seem to be mere property-ascribing statements, natural or non-natural; they express favorable or unfavorable attitudes (and do not merely generate them), recommend, prescribe, and the like. Of course, one could maintain that they do this and also ascribe simple non-natural properties to actions and things, but such a view still involves one in the difficulties mentioned a moment ago. The main point to be made now is that the belief in self-evident ethical truths, and all that goes with it, is so difficult to defend that it seems best to look for some other answer to the problem of justification.

NONCOGNITIVE OR NONDESCRIPTIVIST THEORIES

The third general type of theory of the meaning or nature of ethical judgments has no very satisfactory label. However, it has been called noncognitivist or nondescriptivist because, as against both definists and intuitionists, it holds that ethical judgments are not assertions or statements ascribing properties to (or denying them of) actions, persons, or things, and insists that they have a very different "logic," meaning, or use. It embraces a wide variety of views, some more and others much less extreme.

1. The most extreme of these are a number of views that deny ethical judgments, or at least the most basic ones, to be capable of any kind of

rational or objectively valid justification. On one such view—that of A. J. Ayer—they are simply expressions of emotion much like ejaculations. Saying that killing is wrong is like saying, "Killing, boo!" It says nothing true or false and cannot be justified in any rational way. Rudolf Carnap once took a similar view, except that he interpreted "Killing is wrong" as a command, "Do not kill," rather than as an ejaculation. Bertrand Russell held that moral judgments merely express a certain kind of wish. Many existentialists likewise regard basic ethical judgments, particular or general, as arbitrary commitments or decisions for which no justification can be given.

I should point out here that such irrationalistic views about ethical judgments are not held only by atheistic positivists and existentialists. They are also held by at least some religious existentialists and by other theologians. For example, a theologian who maintains that the basic principles of ethics are divine commands is taking a position much like Carnap's. If he adds that God's commands are arbitrary and cannot be justified rationally, then his position is no less extreme. If he holds that God's commands are, at least in principle, rationally defensible, then his position is like the less extreme ones to be described.

2. C. L. Stevenson's form of the emotive theory is somewhat less extreme than Ayer's. He argues that ethical judgments express the speaker's attitudes and evoke, or seek to evoke, similar attitudes in the hearer. But he realizes that to a very considerable extent our attitudes are based on our beliefs, and so can be reasoned about. For example, I may favor a certain course of action because I believe it has or will have certain results. I will then advance the fact that it has these results as an argument in its favor. But you may argue that it does not have these results, and if you can show this, my attitude may change and I may withdraw my judgment that the course of action in question is right or good. In a sense, you have refuted me. But, of course, this is only because of an underlying attitude on my part of being in favor of certain results rather than others. Stevenson goes on to suggest that our most basic attitudes, and the ethical judgments in which we express them, may not be rooted in beliefs of any kind, in which case they cannot be reasoned about in any way. He is open-minded about this, however, and allows a good deal of room for a kind of argument and reasoning.

3. More recently, from a number of Oxford philosophers and others, we have had still less extreme views. They refuse to regard ethical judgments as mere expressions or evocations of feeling or attitude, as mere commands, or as arbitrary decisions or commitments. Rather, they regard them as evaluations, recommendations, prescriptions, and the like; and they stress the fact that such judgments imply that we are willing to generalize or universalize them and are ready to reason about them, points with which we have agreed. That is, they point out that when we say of something that it is good or right, we imply that there are reasons for our judgment which are

not purely persuasive and private in their cogency. They are even ready to say that such a judgment may be called true or false, though it is very different from "X is yellow" or "Y is to the left of Z." For them ethical judgments are essentially reasoned acts of evaluating, recommending, and prescribing.

The arguments for such theories—the open question argument against definists and the two arguments against intuitionists—we have found to be less conclusive than they are thought to be. To my mind, nevertheless, these theories, or rather the least extreme of them, are on the right track. The kind of account the latter give of the meaning and nature of ethical judgments is acceptable as far as it goes. Such judgments do not simply say that something has or does not have a certain property. Neither are they mere expressions of emotion, will, or decision. They do more than just express or indicate the speaker's attitudes. They evaluate, instruct, recommend, prescribe, advise, and so on; and they claim or imply that what they do is rationally justified or justifiable, which mere expressions of emotion and commands do not do. The more extreme views, therefore, are mistaken as a description of the nature of ethical judgments. Moreover, it is not necessary to agree with them that such judgments cannot be justified in any important sense. They generally assume that if such judgments are not self-evident and cannot be proved inductively or deductively on the basis of empirical or nonempirical facts, as we have seen to be the case, then it follows that they are purely arbitrary. But this does not follow. It may be that this conception of rational justification is too narrow, as I have already intimated in discussing psychological egoism and hedonism. Mill may be right when he says, near the end of Chapter I of *Utilitarianism*,

> We are not . . . to infer that [the acceptance or rejection of an ethical first principle] must depend on blind impulse, or arbitrary choice. There is a larger meaning of the word "proof," in which this question is . . . amenable to it . . . The subject is within the cognizance of the rational faculty; and neither does that faculty deal with it solely in the way of intuition. Considerations may be presented capable of determining the intellect either to give or withhold its assent . . .

Here, Mill is with the less extreme of the recent nondescriptive theories, as against the definists, the intuitionists, and the more extreme nondescriptivists. All of these share the conception of justification as consisting either in self-evidence or in inductive or deductive proof. Only the definists and intuitionists believe that ethical judgments can be justified in one or the other of these ways, while positivists and existentialists deny that ethical judgments can be justified at all. Mill and the less extreme recent philosophers, on the other hand, agree with intuitionists and definists that they can be justified in some rational sense or in some "larger meaning of the word 'proof'," though they have different and various views about the nature of such justification.

At this point, it may help to notice that even such things as "mere" expressions of feeling and commands may be justified or unjustified, rational or irrational. Suppose that A is angry at B, believing B to have insulted him. C may be able to show A that his anger is unjustified, since B has not actually insulted him at all. If A simply goes on being angry, although he no longer has any reason, we should regard his anger as quite irrational. Again, if an officer commands a private to close the door, believing it to be open when it is not, it is reasonable for the private to answer, "But, sir, the door is closed," and it would be quite irrational if the officer were seriously to command the private to close it anyway. Emotions and commands, generally at least, have a background of beliefs and are justified or unjustified, rational or irrational, depending on whether these beliefs themselves are so.

APPROACH TO AN ADEQUATE THEORY

In my opinion, even the less extreme of recent nondescriptivist theories have not gone far enough. They have been too ready to admit a kind of basic relativism after all. They insist that ethical judgments imply the presence of, or at least the possibility of giving, reasons which justify them. But they almost invariably allow or even insist that the validity of these reasons is ultimately relative, either to the individual or to his culture, and therefore, conflicting basic judgments may both be justified or justifiable. Now, it may be that in the end one must agree with this view, but most recent discussions entirely neglect a fact about ethical judgments on which Ewing has long insisted, namely, that they make or somehow imply a claim to be objectively and rationally justified or valid. In other words, an ethical judgment claims that it will stand up under scrutiny by oneself and others in the light of the most careful thinking and the best knowledge, and that rival judgments will not stand up under such scrutiny. Hume makes the point nicely, though only for *moral* judgments:

> The notion of morals implies some sentiment common to all mankind, which recommends the same object to general approbation. . . . When a man denominates another his *enemy*, his *rival*, his *antagonist*, his *adversary*, he is understood to speak the language of self-love, and to express sentiments, peculiar to himself, and arising from his particular circumstance and situation. But when he bestows on any man the epithets of *vicious* or *odious* or *depraved*,

> he then speaks another language, and expresses sentiments, in which he expects all his audience are to concur with him. He must here . . . depart from his private and particular situation, and must choose a point of view, common to himself with others. . . . [4]

And, he must claim, Hume might have added, that anyone else who takes this point of view and from it reviews the relevant facts will come to the same conclusion. In fact, he goes on to suggest that precisely because we need or want a language in which to express, not just sentiments peculiar to ourselves but sentiments in which we expect all men are to concur with us, another language in which we may claim that our sentiments are justified and valid, we had to

> . . . invent a peculiar set of terms, in order to express those universal sentiments of censure or approbation. . . . Virtue and vice become then known; morals are recognized; certain general ideas are framed of human conduct and behavior. . . .

This kind of account of our normative discourse strikes me as eminently wise. It is a language in which we may express our sentiments—approvals, disapprovals, evaluations, recommendations, advice, instructions, prescriptions—and put them out into the public arena for rational scrutiny and discussion, claiming that they will hold up under such scrutiny and discussion and that all our audience will concur with us if they will also choose the same common point of view. That this is so is indicated by the fact that if A makes an ethical judgment about X and then, upon being challenged by B, says, "Well, at least I'm in favor of X," we think he has backed down. He has shifted from the language of public dialogue to that of mere self-revelation. This view recognizes the claim to objective validity on which intuitionists and definists alike insist, but it also recognizes the force of much recent criticism of such views.

RELATIVISM

Against any such view it will be argued, of course, that this claim to be objectively and rationally justified or valid, in the sense of holding up against all rivals through an impartial and informed examination, is simply mistaken and must be given up. This is the contention of the relativist and we must consider it now, although we can do so only briefly.

Actually, we must distinguish at least three forms of relativism. First, there is what may be called *descriptive relativism.* When careful, it does not say merely that the ethical judgments of different people and societies are different. For this would be true even if people and societies agreed in their basic ethical judgments and differed only in their *derivative* ones. What careful descriptive relativism says is that the *basic* ethical beliefs of different people and societies are different and even conflicting. I stress this because the fact that in some primitive societies children believe they should put their parents to death before they get old, whereas we do not, does not prove descriptive relativism. These primitive peoples may believe this because they think their parents will be better off in the hereafter if they enter it while still ablebodied; if this is the case, their ethics and ours are alike in that they rest on the precept that children should do the best they can for their parents. The divergence, then, would be in factual, rather than in ethical, beliefs.

Second, there is *meta-ethical relativism*, which is the view we must consider. It holds that, in the case of basic ethical judgments, there is no objectively valid, rational way of justifying one against another; consequently, two conflicting basic judgments may be equally valid.

The third form of relativism is *normative relativism*. While descriptive relativism makes an anthropological or sociological assertion and meta-ethical relativism a meta-ethical one, this form of relativism puts forward a normative principle: what is right or good for one individual or society is not right or good for another, even if the situations involved are similar, meaning not merely that what is thought right or good by one is not thought right or good by another (this is just descriptive relativism over again), but that what is really right or good in the one case is not so in another. Such a normative principle seems to violate the requirements of consistency and universalization mentioned earlier. We need not consider it here, except to point out that it cannot be justified by appeal to either of the other forms of relativism and does not follow from them. One can be a relativist of either of the other sorts without believing that the same kind of conduct is right for one person or group and wrong for another. One can, for example, believe that everyone ought to treat people equally, though recognizing that not everyone admits this and holding that one's belief cannot be justified.

Our question is about the second kind of relativism. The usual argument used to establish it rests on descriptive relativism. Now, descriptive relativism has not been incontrovertibly established. Some cultural anthropologists and social psychologists have even questioned its truth, for example, Ralph Linton and S. E. Asch. However, to prove meta-ethical relativism one must prove more than descriptive relativism. One must also prove that people's basic ethical judgments would differ and conflict even if they were fully enlightened and shared all the same factual beliefs. It is not enough to show that people's basic ethical judgments are different, for such differences might all be due to differences and incompleteness in their factual beliefs, as in the example of the primitive societies used previously. In this case, it would still be possible to hold that some basic ethical judgments can be justified as valid to everyone, in principle at least, if not in practice.

It is, however, extremely difficult to show that people's basic ethical judgments would still be different even if they were fully enlightened, conceptually clear, shared the same factual beliefs, and were taking the same point of view. To show this, one would have to find clear cases in which all of these conditions are fulfilled and people still differ. Cultural anthropologists do not show us such cases; in all of their cases, there are differences in conceptual understanding and factual belief. Even when one takes two people in the same culture, one cannot be sure that all of the necessary conditions are fulfilled. I conclude, therefore, that meta-ethical relativism has not been proved and, hence, that we need not, in our ethical judgments, give up the claim that they are objectively valid in the sense that they will be sustained by a review by all those who are free, clear-headed, fully informed, and who take the point of view in question.

A THEORY OF JUSTIFICATION

We now have the beginnings of a theory of the meaning and justification of ethical judgments. To go any farther, we must distinguish moral judgments proper from nonmoral normative judgments and say something separately about the justification of each. How can we distinguish moral from other normative judgments? Not by the words used in them, for words like "good" and "right" all have nonmoral as well as moral uses. By the feelings that accompany them? The difficulty in this proposal is that it is hard to tell which feelings are moral except by seeing what judgments they go with. It is often thought that moral judgments are simply whatever judgments we regard as overriding all other normative judgments in case of conflict, but then aesthetic or prudential judgments become moral ones if we take them to have priority over others, which seems paradoxical. It seems to me that what makes some normative judgments moral, some aesthetic, and some prudential is the fact that different points of view are taken in the three cases, and that the point of view taken is indicated by the kinds of reasons that are given. Consider three judgments: (a) I say that you ought to do X and give as the reason the fact that X will help you succeed in business; (b) I say you should do Y and cite as the reason the fact that Y will produce a striking contrast of colors; and (c) I say you should do Z and give as the reason the fact that Z will keep a promise or help someone. Here the reason I give reveals the point of view I am taking and the kind of judgment I am making.

Now let us take up the justification of nonmoral normative judgments. We are interested primarily in judgments of intrinsic value such as were discussed in the previous chapter, for such judgments are relevant to ethics because, through the principle of beneficence, the question of what is good or bad comes to bear on the question of what is right or wrong. Besides, if we know how to justify judgments of intrinsic value, we will know how to justify judgments of extrinsic and inherent value, for judgments of the latter sorts presuppose judgments of the former. It is true, as we have already seen, that we cannot *prove* basic judgments of intrinsic value in any strict sense of proof, but this fact does not mean that we cannot justify them or reasonably claim them to be justified. But how can we do this? By taking what I shall call the evaluative point of view as such, unqualified by any such adjective as "aesthetic," "moral," or "prudential," and then trying to see what judgment we are led to make when we do so, considering the thing in question wholly on the basis of its intrinsic character, not its consequences or conditions. What is it to take the nonmorally evaluative point of view? It is to be free, informed, clear-headed, impartial, willing to universalize; in general, it is to be "calm" and "cool," as Butler would say, in one's consideration of such items as pleasure, knowledge, and love, for the question is simply what it is rational to choose. This is what we tried to do in Chapter 5. If one considers an item in this reflective way and comes out in favor of it, one is rationally justified in judging it to be intrinsically good, even if one cannot prove one's judgment. In doing so, one claims that everyone else who does likewise will concur; and one's judgment is really justified if this claim is correct, which, of course, one can never know for certain. If others who also claim to be calm and cool do not concur, one must reconsider to see if both sides are really taking the evaluative point of view, considering only intrinsic features, clearly understanding one another, and so on. More one cannot do and, if disagreement persists, one may still claim to be right (i.e., that others will concur eventually if . . .); but one must be open-minded and tolerant. In fact, we saw in Chapter 5 that one may have to admit a certain relativity in the ranking of things listed as intrinsically good, although possibly not in the listing itself.

What about the justification of moral judgments? Already in Chapters 2 and 3 we have, in effect, said something about the justification of judgments of right, wrong, and obligation. We argued that a particular judgment essentially entails a general one, so that one cannot regard a particular judgment as justified unless one is also willing to accept the entailed general one, and vice versa. This is true whether we are speaking of judgments of actual or of prima facie duty. We have also seen that judgments of actual duty, whether particular judgments or rules, cannot simply be deduced from the basic principles of beneficence and justice, even with the help of factual premises, since these must be taken as prima facie principles and may conflict on occasion. Thus, we have two question: first, how can we justify judgments of actual duty, general or particular, and second, how can we justify basic principles of prima facie duty? The same answer, however, will do for both. It seems fair to assume that it will also do for the question of justifying judgments of moral value.

First, we must take the moral point of view, as Hume indicated, not that of self-love or aesthetic judgment, nor the more general point of view involved in judgments of intrinsic value. We must also be free, impartial, willing to universalize, conceptually clear, and informed about all possibly relevant facts. Then we are justified in judging that a certain act or kind of action is right, wrong, or obligatory, and in claiming that our judgment is objectively valid, at least as long as no one who is doing likewise disagrees. Our judgment or principle is really justified if it holds up under sustained scrutiny of this sort from the moral point of view on the part of everyone. Suppose we encounter someone who claims to be doing this but comes to a different conclusion. Then we must do our best, through reconsideration and discussion, to see if one of us is failing to meet the conditions in some way. If we can detect no failing on either side and still disagree, we may and I think still must each claim to be correct, for the conditions never are perfectly fulfilled by both of us and one of us may turn out to be mistaken after all. If what was said about relativism is true, we cannot both be correct. But both of us must be open-minded and tolerant if we are to go on living within the moral institution of life and not resort to force or other immoral or nonmoral devices.

If this line of thought is acceptable, then we may say that a basic moral judgment, principle, or code is justified or "true" if it is or will be agreed to by everyone who takes the moral point of view and is clearheaded and logical and knows all that is relevant about himself, mankind, and the universe. Are our own principles of beneficence and justice justified or "true" in this sense? The argument in Chapters 2 and 3 was essentially an attempt to take the moral point of view and from it to review various normative theories and arrive at one of our own. Our principles have not been proved, but perhaps it may be claimed that they will be concurred in by those who try to do likewise. This claim was implicitly made in presenting them. Whether the claim is true or not must wait upon the scrutiny of others.

The fact that moral judgments claim a consensus on the part of others does not mean that the individual thinker must bow to the judgment of the majority in his society. He is not claiming an *actual* consensus, he is claiming that in the end—which never comes or comes only on the Day of Judgment—his position will be concurred in by those who freely and clear-headedly review the relevant facts from the moral point of view. In other words, he is claiming an *ideal* consensus that transcends majorities and actual societies. One's society and its code and institutions may be wrong. Here enters the autonomy of the moral agent—he must take the moral point of view and must claim an eventual consensus with others who do so, but he must judge for himself. He may be mistaken, but, like Luther, he cannot do otherwise. Similar remarks hold for one who makes nonmoral judgments.

THE MORAL POINT OF VIEW

What is the moral point of view? This is a crucial question for the view we have suggested. It is also one on which there has been much controversy lately. According to one theory, one is taking the moral point of view if and only if one is willing to universalize one's maxims. Kant would probably accept this if he were alive. But I pointed out that one may be willing to universalize from a prudential point of view; and also that what one is willing to universalize is not necessarily a moral rule. Other such formal characterizations of the moral point of view have been proposed. A more plausible characterization to my mind, however, is that of Kurt Baier. He holds that one is taking the moral point of view if one is not being egoistic, one is doing things on principle, one is willing to universalize one's principles, and in doing so one considers the good of everyone alike.[5]

Hume thought that the moral point of view was that of sympathy, and it seems to me he was on the right wavelength. I have already argued that the point of view involved in a judgment can be identified by the kind of reason that is given for the judgment when it is made or if it is challenged. Then the moral point of view can be identified by determining what sorts of facts are reasons for moral judgments or moral reasons. Roughly following Hume, I now want to suggest that moral reasons consist of facts about what actions, dispositions, and persons do to the lives of sentient beings, including beings other than the agent in question, and that the moral point of view is that which is concerned about such facts. My own position, then, is that one is taking the moral point of view if and only if (a) one is making normative judgments about actions, desires, dispositions, intentions, motives, persons, or traits of character; (b) one is willing to universalize one's judgments; (c) one's reasons for one's judgments consist of facts about what the things judged do to the lives of sentient beings in terms of promoting or distributing nonmoral good and evil; and (d) when the judgment is about oneself or one's own actions, one's reasons include such facts about what one's own actions and dispositions do to the lives of other sentient beings as such, if others are affected. One has a morality or moral action-guide only if and insofar as one makes normative judgments from this point of view and is guided by them.

WHY BE MORAL?

Another problem that remains has been mentioned before. Why should we be moral? Why should we take part in the moral institution of life? Why should we adopt the moral point of view? We have already seen that the question, "Why should . . . ?" is ambiguous, and may be a request either for motivation or for justification. Here, then, one may be asking for (1) the motives for doing what is morally right, (2) a justification for doing what is morally right, (3) motivation for adopting the moral point of view and otherwise subscribing to the moral institution of life, or (4) a justification of morality and the moral point of view. It is easy to see the form an answer to a request for (1) and (3) must take; it will consist in pointing out the various prudential and non-prudential motives for doing what is right or for participating in the moral institution of life. Most of these are familiar or readily thought of and need not be detailed here. A request for (2) might be taken as a request for a *moral* justification for doing what is right. Then, the answer is that doing what is morally right does not need a justification, since the justification has already been given in showing that it is right. On this interpretation, a request for (2) is like asking, "Why morally ought I to do what is morally right?" A request for (2) may also, however, be meant as a demand for a nonmoral justification of doing what is morally right; then, the answer to it will be like the answer to a request for (4). For a request for (4), being a request for reasons for subscribing to the moral way of thinking, judging, and living, must be a request for a nonmoral justification of morality. What will this be like?

There seem to be two questions here. First, why should *society* adopt such an institution as morality? Why should it foster such a system for the guidance of conduct in addition to convention, law, and prudence? To this the answer seems clear. The conditions of a satisfactory human life for people living in groups could hardly obtain otherwise. The alternatives would seem to be either a state of nature in which all or most of us would be worse off than we are, even if Hobbes is wrong in thinking that life in such a state would be "solitary, poor, nasty, brutish, and short"; or a leviathan civil state more totalitarian than any yet dreamed of, one in which the laws would cover all aspects of life and every possible deviation by the individual would be closed off by an effective threat of force.

The other question has to do with the nonmoral reasons (not just motives) there are for an *individual's* adopting the moral way of thinking and living. To some extent, the answer has just been given, but only to some extent. For on reading the last paragraph an individual might say, "Yes. This shows that society requires morality and even that it is to my advantage to have others adopt the moral way of life. But it does not show that I should adopt it, and certainly not that I should *always* act according to it. And it is no use arguing on moral grounds that I should. I want a nonmoral justification for thinking I should." Now, if this means that he wants to be shown that it is always to his advantage—that is, that his life will invariably be better or, at least, not worse in the prudential sense of better and worse—if he thoroughly adopts the moral way of life, then I doubt that his demand can always be met. Through the use of various familiar arguments, one can show that the moral way of life is likely to be to his advantage, but it must be admitted in all honesty that one who takes the moral road may be called upon to make a sacrifice and, hence, may not have as good a life in the nonmoral sense as he would otherwise have had.

The point made at the end of Chapter 5 must be recalled here, namely, that morally good or right action is one kind of excellent activity and hence is a prime candidate for election as part of any good life, especially since it is a kind of excellent activity of which all normal people are capable. It does seem to me that this is an important consideration in the answer to our present question. Even if we add it to the usual arguments, however, we still do not have a conclusive proof that every individual should, in the nonmoral sense under discussion, always do the morally excellent thing. For, as far as I can see, from a prudential point of view, some individuals might have nonmorally better lives if they sometimes did what is not morally excellent, for example in cases in which a considerable self-sacrifice is morally required. A TV speaker once said of his subject, "He was too good for his good," and it seems to me that this may sometimes be true.

It does not follow that one cannot justify the ways of morality to an individual, although it may follow that one cannot justify morality to some individuals. For nonmoral justification is not necessarily egoistic or prudential. If A asks B why he, A, should be moral, B may reply by asking A to try to decide in a rational way what kind of a life he wishes to live or what kind of a person he wishes to be. That is, B may ask A what way of life A would choose if he were to choose rationally, or in other words, freely, impartially, and in full knowledge of what it is like to live the various alternative ways of life, including the moral one. B may then be able to convince A, when he is calm and cool in this way, that the way of life he prefers, all things considered, includes the moral way of life. If so, then he has justified the moral way of life to A. A may even, when he considers matters in such a way, prefer a life that includes self-sacrifice on his part.

Of course, A may refuse to be rational, calm, and cool. He may retort, "But why should I be rational?" However, if this was his posture in originally asking for justification, he had no business asking for it. For one can only ask for justification if one is willing to be rational. One cannot consistently ask for reasons unless one is ready to accept reasons of some sort. Even in asking, "Why should I be rational?" one is implicitly committing oneself to rationality, for such a commitment is part of the connotation of the word "should."

What kind of a life A would choose if he were fully rational and knew all about himself and the world will, of course, depend on what sort of a person he is (and people are different), but if psychological egoism is not true of any of us, it may always be that A would then choose a way of life that would be moral. As Bertrand Russell once wrote:

> We have wishes which are not purely personal . . . The sort of life that most of us admire is one which is guided by large, impersonal desires . . . Our desires are, in fact, more general and less purely selfish than many moralists imagine . . .[6]

Perhaps A has yet one more question: Is society justified in demanding that I adopt the moral way of life, and in blaming and censuring me if I do not?" But this is a moral question; and A can hardly expect it to be allowed that society is justified in doing this to A only if it can show that doing so is to A's advantage. However, if A is asking whether society is morally justified in requiring of him at least a certain minimal subscription to the moral institution of life, then the answer surely is that the society sometimes is justified in this, as Socrates argued in the *Crito*. But society must be careful here. For it is itself morally required to respect the individual's autonomy and liberty, and in general to treat him justly; and it must remember that morality is made to minister to the good lives of individuals and not to interfere with them any more than is necessary. Morality is made for man, not man for morality.

NOTES

1. *Realms of Value* (Cambridge, Mass.: Harvard University Press, 1954), pp. 3, 107, 109. See selections in Frankena and Granrose, Chap. VI.
2. Most writers today use "naturalism" to cover all kinds of definism.
3. *Ethics* (New York: The Century Co., 1928), pp. 109, 410–11.
4. *An Enquiry into the Principles of Morals*, pp. 113–14.

5. *The Moral Point of View* (New York: Random House, 1965), Chap. 5.
6. *Religion and Science* (New York: Henry Holt and Co., 1935), pp. 252–54.

Part Two: Educational Aims

Introduction

Johann Herbart was of the opinion that education looks to ethics for its ends and to psychology for the means to those ends.[1] That is an oversimplification, no doubt, but the fact remains that educational ends or aims do entail value judgments, and the latter are a vital part of ethics.

In any case, John Dewey, as the one-time chairman of the Department of Philosophy, Psychology, and Pedagogy at the University of Chicago, was well aware of the interrelationships between the three disciplines. The following chapter from *Democracy and Education* reflects that awareness and touches on a number of ideas central to Dewey's educational thought. We find Dewey arguing, for example, "that the object and reward of learning is continued capacity for growth." (In an earlier chapter he takes the controversial position that since life is growth and education is life, then education is growth and has no end other than growth.)[2] He also insists that ends and means cannot be separated, and that an educational aim "must be an outgrowth of existing conditions." Thus Dewey objects to "externally imposed," and overly abstract aims.

Other familiar Deweyan concepts encountered in this chapter include his notions of mind, intelligent behavior, the significance of anticipated consequences as an indicator of intelligence, and the educational import of individual differences.

The selection by Sidney Hook is taken from his influential book, *Education for Modern Man*, and is to some extent similar in outlook to Dewey's views on educational aims. Both writers see themselves as "experimentalists" in formulating educational ends, for example, and both include growth among those ends. Professor Hook points out, incidentally, that Dewey's (and Hook's) notion of growth is inseparable from his democratic social and educational philosophy.

Taking "an experimentalist approach to the question of educational ends" (which he contrasts with metaphysical and theological approaches), Hook lists growth, democracy, and intelligence as the ends of education, and stresses the close relationship between the three. These three ends, Hook suggests, are consistent with man's nature as a social organism that exhibits certain biologically and culturally influenced patterns of behavior. In justifying his selection of the above ends, Professor Hook provides a clear account of the grounds for his choices and in the process exhibits his experimentalist method in action.

In this chapter Professor Hook also explains how it comes about that those who subscribe to similar educational aims often differ sharply over their application. He views conflicting values and disagreement over the meaning of ethical language as the two chief sources of differences at the level of practice.

In the third selection, William Frankena construes educational aims or ends as certain dispositions, "which it is the business of education to promote." He focuses on the clusters of dispositions advocated by three of the leading schools of contemporary philosophy: Deweyan pragmatism, analytical philosophy, and existentialism. Dewey, Frankena points out, was concerned with promoting "reflective thinking, scientific intelligence or experimental inquiry." The analysts, on the other hand, emphasize "clarity, consistency, [and] rigor of thought." And finally, the existentialists stress "authenticity," "commitment," "responsibility," and the like.

Professor Frankena suggests that although the three philosophies are often in disagreement with one another, the combination of dispositions they favor may serve as worthy educational aims. But they need, in Frankena's view, to be supplemented by additional dispositions, such as those relating to various forms of knowledge, moral and political values, and aesthetic appreciation. The essay closes with a brief discussion of still other dispositions that may be appropriate for private education.

NOTES

1. Harold B. Dunkel, *Herbart and Herbartianism: An Educational Ghost Story* (Chicago: The University of Chicago Press, 1970), p. 83.
2. John Dewey, *Democracy and Education* (New York: Macmillan paperback ed., 1961). Chapter 4.

4. Aims in Education

John Dewey

1. THE NATURE OF AN AIM

The account of education given in our earlier chapters virtually anticipated the results reached in a discussion of the purport of education in a democratic community. For it assumed that the aim of education is to enable individuals to continue their education—or that the object and reward of learning is continued capacity for growth. Now this idea cannot be applied to *all* the members of a society except where intercourse of man with man is mutual, and except where there is adequate provision for the reconstruction of social habits and institutions by means of wide stimulation arising from equitably distributed interests. And this means a democratic society. In our search for aims in education, we are not concerned, therefore, with finding an end outside of the educative process to which education is subordinate. Our whole conception forbids. We are rather concerned with the contrast which exists when aims belong within the process in which they operate and when they are set up from without. And the latter state of affairs must obtain when social relationships are not equitably balanced. For in that case, some portions of the whole social group will find their aims determined by an external dictation; their aims will not arise from the free growth of their own experience, and their nominal aims will be means to more ulterior ends of others rather than truly their own.

Our first question is to define the nature of an aim so far as it falls within an activity, instead of being furnished from without. We approach the definition by a contrast of mere *results* with *ends*. Any exhibition of energy has results. The wind blows about the sands of the desert; the position of the grains is changed. Here is a result, an effect, but not an *end*. For there is nothing in the outcome which completes or fulfills what went before it. There is mere spatial redistribution. One state of affairs is just as good as any other. Consequently there is no basis upon which to select an earlier state of affairs as a beginning, a later as an end, and to consider what intervenes as a process of transformation and realization.

Consider for example the activities of bees in contrast with the changes in the sands when the wind blows them about. The results of the bees' actions may be called ends not because they are designed or consciously intended, but because they are true terminations or completions of what has preceded. When the bees gather pollen and make wax and build cells, each step prepares the way for the next. When cells are built, the queen lays eggs in them; when eggs are laid, they are sealed and bees brood them and keep them at a temperature required to hatch them. When they are hatched, bees feed the young till they can take care of themselves. Now we are so familiar with such facts, that we are apt to dismiss them on the ground that life and instinct are a kind of miraculous thing anyway. Thus we fail to note what the essential characteristic of the event is; namely, the significance of the temporal place and order of each element; the way each prior event leads into its successor while the successor takes up what is furnished and utilizes it for some other stage, until we arrive at the end, which, as it were, summarizes and finishes off the process.

Since aims relate always to results, the first thing to look to when it is a question of aims, is whether the work assigned possesses intrinsic continuity. Or is it a mere serial aggregate of acts, first doing one thing and then another? To talk about an educational aim when approximately each act of a pupil is dictated by the teacher, when the only order in the sequence of his acts is that which comes from the assignment of lessons and the giving of directions by another, is to talk nonsense. It is equally fatal to an aim to permit capricious or discontinuous action in the name of spontaneous self-expression. An aim implies an orderly and ordered activity, one in which the order consists in the progressive completing of a process. Given an activity having a time span and cumulative growth within the time succession, an aim means foresight in advance of the end or possible termination. If bees anticipated the consequences of their activity, if they perceived their end in an imaginative foresight, they would have the primary element in an aim. Hence it is nonsense to talk about the aim of education—or any other undertaking—where conditions do not permit of foresight of results, and do not stimulate a person to look ahead to see what the outcome of a given activity is to be.

In the next place the aim as a foreseen end gives direction to the activity; it is not an idle view of a mere spectator, but influences the steps taken to reach the end. The foresight functions in three ways. In the first place, it involves careful observation of the given conditions to see what are the means available for reaching the end, and to discover the hindrances in the way. In the second place, it suggests the proper order or sequence in the use of means. It facilitates an economical selection and arrangement. In the third place, it makes choice of alternatives possible. If we can predict the outcome of acting this way or that, we can then compare the value of the two courses of action; we can pass judgment upon their relative desirability. If we know that stagnant water breeds mosquitoes and that they are likely to carry disease, we can, disliking that anticipated result, take steps to avert it. Since we do not anticipate results as mere intellectual onlookers, but as persons concerned in the outcome, we are partakers in the process which produces the result. We intervene to bring about this result or that.

Of course these three points are closely connected with one another. We can definitely foresee results only as we make careful scrutiny of present conditions, and the importance of the outcome supplies the motive for observations. The more adequate our observations, the more varied is the scene of conditions and obstructions that presents itself, and the more numerous are the alternatives between which choice may be made. In turn, the more numerous the recognized possibilities of the situation, or alternatives of action, the more meaning does the chosen activity possess, and the more flexibly controllable is it. Where only a single outcome has been thought of, the mind has nothing else to think of; the meaning attaching to the act is limited. One only steams ahead toward the mark. Sometimes such a narrow course may be effective. But if unexpected difficulties offer themselves, one has not as many resources at command as if he had chosen the same line of action after a broader survey of the possibilities of the field. He cannot make needed readjustments readily.

The net conclusion is that acting with an aim is all one with acting intelligently. To foresee a terminus of an act is to have a basis upon which to observe, to select, and to order objects and our own capacities. To do these things means to have a mind—for mind is precisely intentional purposeful activity controlled by perception of facts and their relationships to one another. To have a mind to do a thing is to foresee a future possibility; it is to have a plan for its accomplishment; it is to note the means which make the plan capable of execution and the obstructions in the way,—or, if it is really a *mind* to do the thing and not a vague aspiration—it is to have a plan which takes account of resources and difficulties. Mind is capacity to refer present conditions to future results, and future consequences to present conditions. And these traits are just what is meant by having an aim or a purpose. A man is stupid or blind or unintelligent—lacking in mind—just in the degree in which in any activity he does not know what he is about, namely, the probable consequences of his acts. A man is imperfectly intelligent when he contents himself with looser guesses about the outcome than is needful, just taking a chance with his luck, or when he forms plans apart from study of the actual conditions, including his own capacities. Such relative absence of mind means to make our feelings the measure of what is to happen. To be intelligent we must "stop, look, listen" in making the plan of an activity.

To identify acting with an aim and intelligent activity is enough to show its value—its function in experience. We are only too given to making an entity out of the abstract noun "consciousness." We forget that it comes from the adjective "conscious." To be conscious is to be aware of what we are about; conscious signifies the deliberate, observant, planning traits of activity. Consciousness is nothing which we have which gazes idly on the scene around one or which has impressions made upon it by physical things; it is a name for the purposeful quality of an activity, for the fact that it is directed by an aim. Put the other way about, to have an aim is to act with meaning, not like an automatic machine; it is to *mean* to do something and to perceive the meaning of things in the light of that intent.

2. THE CRITERIA OF GOOD AIMS

We may apply the results of our discussion to a consideration of the criteria involved in a correct establishing of aims. (1) The aim set up must be an outgrowth of existing conditions. It must be based upon a consideration of what is already going on; upon the resources and difficulties of the situation. Theories about the proper end of our activities—educational and moral theories—often violate this principle. They assume ends lying *outside* our activities; ends foreign to the concrete makeup of the situation; ends which issue from some outside source. Then the problem is to bring our activities to bear upon the realization of these externally supplied ends. They are something for which we *ought* to act. In any case such "aims" limit intelligence; they are not the expression of mind in foresight, observation, and choice of the better among alternative possibilities. They limit intelligence because, given ready-made, they must be imposed by some authority external to intelligence, leaving to the latter nothing but a mechanical choice of means.

(2) We have spoken as if aims could be completely formed prior to the attempt to realize them. This impression must now be qualified. The aim as it first emerges is a mere tentative sketch. The act of striving to realize it tests its worth. If it suffices to direct activity successfully, nothing more is required, since its whole function is to set a mark in advance; and at times a mere hint may suffice. But usually—at least in complicated situations—acting upon it brings to light conditions which had been overlooked. This

calls for revision of the original aim; it has to be added to and subtracted from. An aim must, then, be *flexible;* it must be capable of alteration to meet circumstances. An end established externally to the process of action is always rigid. Being inserted or imposed from without, it is not supposed to have a working relationship to the concrete conditions of the situation. What happens in the course of action neither confirms, refutes, nor alters it. Such an end can only be insisted upon. The failure that results from its lack of adaptation is attributed simply to the perverseness of conditions, not to the fact that the end is not reasonable under the circumstances. The value of a legitimate aim, on the contrary, lies in the fact that we can use it to change conditions. It is a method for dealing with conditions so as to effect desirable alterations in them. A farmer who should passively accept things just as he finds them would make as great a mistake as he who framed his plans in complete disregard of what soil, climate, etc., permit. One of the evils of an abstract or remote external aim in education is that its very inapplicability in practice is likely to react into a haphazard snatching at immediate conditions. A good aim surveys the present state of experience of pupils, and forming a tentative plan of treatment, keeps the plan constantly in view and yet modifies it as conditions develop. The aim, in short, is experimental, and hence constantly growing as it is tested in action.

(3) The aim must always represent a freeing of activities. The term *end in view* is suggestive, for it puts before the mind the termination or conclusion of some process. The only way in which we can define an activity is by putting before ourselves the objects in which it terminates—as one's aim in shooting is the target. But we must remember that the *object* is only a mark or sign by which the mind specifies the *activity* one desires to carry out. Strictly speaking, not the target but *hitting* the target is the end in view; one *takes* aim by means of the target, but also by the sight on the gun. The different objects which are thought of are means of *directing* the activity. Thus one aims at, say, a rabbit; what he wants is to shoot straight: a certain kind of activity. Or, if it is the rabbit he wants, it is not rabbit apart from his activity, but as a factor in activity; he wants to eat the rabbit, or to show it as evidence of his marksmanship—he wants to do something with it. The doing with the thing, not the thing in isolation, is his end. The object is but a phase of the active end,—continuing the activity successfully. This is what is meant by the phrase, used above, "freeing activity."

In contrast with fulfilling some process in order that activity may go on, stands the static character of an end which is imposed from without the activity. It is always conceived of as fixed; it is *something* to be attained and possessed. When one has such a notion, activity is a mere unavoidable means to something else; it is not significant or important on its own account. As compared with the end it is but a necessary evil; something which must be gone through before one can reach the object which is alone worth while.

In other words, the external idea of the aim leads to a separation of means from end, while an end which grows up within an activity as plan for its direction is always both ends and means, the distinction being only one of convenience. Every means is a temporary end until we have attained it. Every end becomes a means of carrying activity further as soon as it is achieved. We call it end when it marks off the future direction of the activity in which we are engaged; means when it marks off the present direction. Every divorce of end from means diminishes by that much the significance of the activity and tends to reduce it to a drudgery from which one would escape if he could. A farmer has to use plants and animals to carry on his farming activities. It certainly makes a great difference to his life whether he is fond of them, or whether he regards them merely as means which he has to employ to get something else in which alone he is interested. In the former case, his entire course of activity is significant; each phase of it has its own value. He has the experience of realizing his end at every stage; the postponed aim, or end in view, being merely a sight ahead by which to keep his activity going fully and freely. For if he does not look ahead, he is more likely to find himself blocked. The aim is definitely a *means* of action as is any other portion of an activity.

3. APPLICATIONS IN EDUCATION

There is nothing peculiar about educational aims. They are just like aims in any directed occupation. The educator, like the farmer, has certain things to do, certain resources with which to do, and certain obstacles with which to contend. The conditions with which the farmer deals, whether as obstacles or resources, have their own structure and operation independently of any purpose of his. Seeds sprout, rain falls, the sun shines, insects devour, blight comes, the seasons change. His aim is simply to utilize these various conditions; to make his activities and their energies work together, instead of against one another. It would be absurd if the farmer set up a purpose of farming, without any reference to these conditions of soil, climate, characteristic of plant growth, etc. His purpose is simply a foresight of the consequences of his energies connected with those of the things about him, a foresight used to direct his movements from day to day. Foresight of possible consequences leads to more careful and extensive observation of the nature and performances of the things he had to do with, and to laying out a plan—that is, of a certain order in the acts to be performed.

It is the same with the educator, whether parent or teacher. It is as absurd for the latter to set up his "own" aims as the proper objects of the growth of the children as it would be for the farmer to set up an ideal of farming irrespective of conditions. Aims mean acceptance of responsibility for the observations, anticipations, and arrangements required in carrying on a function—whether farming or educating. Any aim is of value so far as it assists observation, choice, and planning in carrying on activity from moment to moment and hour to hour; if it gets in the way of the individual's own common sense (as it will surely do if imposed from without or accepted on authority) it does harm.

And it is well to remind ourselves that education as such has no aims. Only persons, parents, and teachers, etc., have aims, not an abstract idea like education. And consequently their purposes are indefinitely varied, differing with different children, changing as children grow and with the growth of experience on the part of the one who teaches. Even the most valid aims which can be put in words will, as words, do more harm than good unless one recognizes that they are not aims, but rather suggestions to educators as to how to observe, how to look ahead, and how to choose in liberating and directing the energies of the concrete situations in which they find themselves. As a recent writer has said: "To lead this boy to read Scott's novels instead of old Sleuth's stories; to teach this girl to sew; to root out the habit of bullying from John's make-up; to prepare this class to study medicine,—these are samples of the millions of aims we have actually before us in the concrete work of education."

Bearing these qualifications in mind, we shall proceed to state some of the characteristics found in all good educational aims. (1) An educational aim must be founded upon the intrinsic activities and needs (including original instincts and acquired habits) of the given individual to be educated. The tendency of such an aim as preparation is, as we have seen, to omit existing powers, and find the aim in some remote accomplishment or responsibility. In general, there is a disposition to take considerations which are dear to the hearts of adults and set them up as ends irrespective of the capacities of those educated. There is also an inclination to propound aims which are so uniform as to neglect the specific powers and requirements of an individual, forgetting that all learning is something which happens to an individual at a given time and place. The larger range of perception of the adult is of great value in observing the abilities and weaknesses of the young, in deciding what they may amount to. Thus the artistic capacities of the adult exhibit what certain tendencies of the child are capable of; if we did not have the adult achievements we should be without assurance as to the significance of the drawing, reproducing, modeling, coloring activities of childhood. So if it were not for adult language, we should not be able to see the import of the babbling impulses of infancy. But it is one thing to use adult accomplishments as a context in which to place and survey the doings of childhood and youth; it is quite another to set them up as a fixed aim without regard to the concrete activities of those educated.

(2) An aim must be capable of translation into a method of cooperating with the activities of those undergoing instruction. It must suggest the kind of environment needed to liberate and to organize *their* capacities. Unless it lends itself to the construction of specific procedures, and unless these procedures test, correct, and amplify the aim, the latter is worthless. Instead of helping the specific task of teaching, it prevents the use of ordinary judgment in observing and sizing up the situation. It operates to exclude recognition of everything except what squares up with the fixed end in view. Every rigid aim just because it is rigidly given seems to render it unnecessary to give careful attention to concrete conditions. Since it *must* apply anyhow, what is the use of noting details which do not count?

The vice of externally imposed ends has deep roots. Teachers receive them from superior authorities; these authorities accept them from what is current in the community. The teachers impose them upon children. As a first consequence, the intelligence of the teacher is not free; it is confined to receiving the aims laid down from above. Too rarely is the individual teacher so free from the dictation of authoritative supervisor, textbook on methods, prescribed course of study, etc., that he can let his mind come to close quarters with the pupil's mind and the subject matter. This distrust of the teacher's experience is then reflected in lack of confidence in the responses of pupils. The latter receive their aims through a double or treble external imposition, and are constantly confused by the conflict between the aims which are natural to their own experience at the time and those in which they are taught to acquiesce. Until the democratic criterion of the intrinsic significance of every growing experience is recognized, we shall be intellectually confused by the demand for adaptation to external aims.

(3) Educators have to be on their guard against ends that are alleged to be general and ultimate. Every activity, however specific, is, of course, general in its ramified connections, for it leads out indefinitely into other things. So far as a general idea makes us more alive to these connections, it cannot be too general. But "general" also means "abstract," or detached from all specific context. And such abstractness means remoteness, and throws us back, once more, upon teaching and learning as mere means of getting ready for an end disconnected from the means. That education is literally and all the time its own reward means that no alleged study or discipline is educative unless it is worth while in its own immediate having. A truly general aim broadens the outlook; it stimulates one to take more consequences (connections) into account. This means a wider and more flexible observation of means. The more interacting forces, for example, the farmer

takes into account, the more varied will be his immediate resources. He will see a greater number of possible starting places, and a greater number of ways of getting at what he wants to do. The fuller one's conception of possible future achievements, the less his present activity is tied down to a small number of alternatives. If one knew enough, one could start almost anywhere and sustain his activities continuously and fruitfully.

Understanding then the term general or comprehensive aim simply in the sense of a broad survey of the field of present activities, we shall take up some of the larger ends which have currency in the educational theories of the day, and consider what light they throw upon the immediate concrete and diversified aims which are always the educator's real concern. We premise (as indeed immediately follows from what has been said) that there is no need of making a choice among them or regarding them as competitors. When we come to act in a tangible way we have to select or choose a particular act at a particular time, but any number of comprehensive ends may exist without competition, since they mean simply different ways of looking at the same scene. One cannot climb a number of different mountains simultaneously, but the views had when different mountains are ascended supplement one another: they do not set up incompatible, competing worlds. Or, putting the matter in a slightly different way, one statement of an end may suggest certain questions and observations, and another statement another set of questions, calling for other observations. Then the more general ends we have, the better. One statement will emphasize what another slurs over. What a plurality of hypotheses does for the scientific investigator, a plurality of stated aims may do for the instructor.

SUMMARY

An aim denotes the result of any natural process brought to consciousness and made a factor in determining present observation and choice of ways of acting. It signifies that an activity has become intelligent. Specifically it means foresight of the alternative consequences attendant upon acting in a given situation in different ways, and the use of what is anticipated to direct observation and experiment. A true aim is thus opposed at every point to an aim which is imposed upon a process of action from without. The latter is fixed and rigid; it is not a stimulus to intelligence in the given situation, but is an externally dictated order to do such and such things. Instead of connecting directly with present activities, it is remote, divorced from the means by which it is to be reached. Instead of suggesting a freer and better balanced activity, it is a limit set to activity. In education, the currency of these externally imposed aims is responsible for the emphasis put upon the notion of preparation for a remote future and for rendering the work of both teacher and pupil mechanical and slavish.

5. The Ends of Education

Sidney Hook

> *"It is true that the aim of education is development of individuals to the utmost of their potentialities. But this statement in isolation leaves unanswered the question as to what is the measure of the development. A society of free individuals in which all, through their own work, contribute to the liberation and enrichment of the lives of others, is the only environment in which any individual can really grow normally to his* full *stature."*
>
> John Dewey

Everyone who makes an intelligent educational decision—whether as student, teacher, parent, or citizen—must sooner or later be able to justify it by reference to what he conceives the ends of education to be. These ends do not alone determine what should be done, but without them action has no focus. It is a matter of form, of imitation or routine.

Where the ends of education are explicitly stated, as they often are in current discussion, there seems to be a wide agreement about them. Differences appear just as soon as we ask about ends the same question we asked about decisions: How are *they* derived and justified? Similarly, there is more agreement about the phrasing of the ends of education than about their concrete meaning in any specific cultural context. I shall try to show that conflicting interpretations of the meaning of educational ends are significantly associated with the different ways in which these ends are derived and applied.

It is not difficult to draw up a list of educational ends to which most educators who are not open apologists for a political or religious church will subscribe independently of their philosophical allegiance.

Reprinted with permission of the author from *Education for Modern Man* (New York: Alfred A. Knopf, 1946, © 1963 by Sidney Hook), pp. 54–67.

(1) Education should aim to develop the powers of critical, independent thought.

(2) It should attempt to induce sensitiveness of perception, receptiveness to new ideas, imaginative sympathy with the experiences of others.

(3) It should produce an awareness of the main streams of our cultural, literary, and scientific traditions.

(4) It should make available important bodies of knowledge concerning nature, society, ourselves, our country, and its history.

(5) It should strive to cultivate an intelligent loyalty to the ideals of the democratic community and to deepen understanding of the heritage of freedom and the prospects of its survival.

(6) At some level, it should equip young men and women with the general skills and techniques and the specialized knowledge which, together with the virtues and aptitudes already mentioned, will make it possible for them to do some productive work related to their capacities and interests.

(7) It should strengthen those inner resources and traits of character which enable the individual, when necessary, to stand alone.

Not only can the seven liberal arts in their modern version be derived from these ends, but we can take them collectively as defining the aim of a liberal education. Any individual in whom the qualities and capacities necessary to achieve these ends have been liberated has received a liberal education. Why, then, should controversy be so rife? After all, if these ends of education are granted, it should not be an insuperable task to decide which specific course of study in a determinate time and place will best realize them. Yet despite the enormous amount of experimental data compiled by educational psychologists, the conflict of schools and philosophies continues unabated. If anything, it has grown more bitter in recent years.

The situation is not unique in education. In the realm of morals, too, we can observe precisely the same thing. Everyone believes, or says he believes, in truth, justice, loyalty, honor, dignity. Yet the strife of moral systems and the diversity of moral judgments in concrete situations, where the same formal values are invoked, is even more conspicuous than in education. In part, the same reason accounts for differences in both moral and educational judgments. Values or goods are plural in morals, just as ends in education are plural. They conflict not only with the values, goods, and ends that are rejected but to some extent among themselves. Two parties to a dispute may both profess allegiance to the ideals of justice *and* happiness or to the goods of security *and* adventure. But they may evaluate them differently, and assign them different weights when faced by the necessity of choice. Similarly, although different schools of education subscribe to critical intelligence *and* loyalty, natural piety for one's traditions *and* independent exploration of new modes of thought, they may be worlds apart in their practical judgments because they accent differently the values they hold in common. They can reach a consensus only insofar as they both submit to a *common method* of resolving conflicts of value in specific situations. But it is at the point of method, i.e., the process by which ideals are themselves derived and evaluated, that they fundamentally divide.

There is another basic reason why the profession of common ends in a common situation is no assurance of agreement. The same words may actually mean different things to those who use them. Anyone who has read Hitler's *Mein Kampf* will find that he invokes many of the ideals of his democratic opponents. He talks about justice, honesty, dignity, craftsmanship, discipline, and willingness to sacrifice. In one passage he asserts that "the importance of the Person" is the distinguishing characteristic of the Nazi philosophy of life which "therefore makes the individual the pillar of the entire edifice."[1] The terms "reason," "freedom," "order," and "discipline" appear in the writings of neo-Thomists, absolute idealists, and experimental naturalists. But they do not mean the same thing. Were one to judge educators only by their *language* in discussing ultimate educational goals, there would be little ground for suspecting the presence of profound differences among them. The situation is closely analogous in political philosophy. Do not all political groups in this country declare their staunch support of the ideals expressed in the Declaration of Independence? Who does not *call* himself a democrat these days? Even Fascists and Communists adorn the chains by which they bind the bodies and spirits of their subjects with flowers of democratic rhetoric.

How, then, do we know when those who accept the same ideals have a common referent or meaning? Roughly, only when these words are conjoined with, or lead to, common behavior or a program of action culminating in common behavior, in a series of common historical situations. This is the prescript of crude common sense as well as of refined scientific method. Indeed, we sometimes come to the conclusion that despite the use of different words people mean the same thing because the programs and behavior to which the words lead are virtually identical. No understanding between human beings is possible without symbols; but the symbols do not have to be verbal. Although it would be extremely difficult, in principle it would not be impossible for human beings to understand each other—on a rather primitive level to be sure—if they could not employ words. But without reference to some kind of co-operative bodily behavior, actual or prospective, remembered or imagined, no matter how long we spoke with one another there would be no assurance of mutual understanding. Even gods and angels have to intrude in the natural order to communicate with men.

The concrete application of these observations to educational ends is twofold. The function of these general ends is not to serve as a rhetorical prologue to programs of education—which is their purpose in the catalogues of most liberal arts colleges—but, first, to suggest a multiplicity of specific ends in the details of instruction. There should be an intimate relation between the objectives of different studies and the underlying ends. Second, these ends must interpenetrate the methods and means of instruction. Nothing is more familiar than the disparity between professed goals and daily practice. Educational ideals operate only insofar as they are continuous with *what* subjects are studied and *how* they are studied. Their true meaning becomes apparent in educational practices and institutions. When these change, the meaning of the ideals changes even when the same words are retained. A hypnotic fixation on words or labels alone creates an insensitiveness, and sometimes an indifference, to the varied contents they conceal.

The most general aims of institutional education in any community are identical with the most general aims of moral (or immoral) action in that community. When we disapprove of the aims of an educational system, and state what they *should* be, we are also indicating, to the extent that they are educationally relevant, what the aims of the good life should be. How, then, do we determine what the aims of education or the good life should be?

There are two generic ways of reaching what are sometimes called "the ultimate" ends of education. One relies on an immediate, self-certifying *intuition* of the nature of man; the other on the observation of the consequences of different proposals of treating man. The first is essentially theological and metaphysical; the second is experimental and scientific.

When they are intelligently formulated both approaches recognize that the ends of education are relevant to the nature of man. But a world of difference separates their conception of the nature of man. The religious or metaphysical approach seeks to deduce what men *should* be from what they *are*. And what they are can only be grasped by an intuition of their "essential" nature. Whatever the differences between Aristotle, Aquinas, and Rousseau on other points—and they are vast—all assert that from the true nature of man the true nature of education follows logically. If we know what man is, then we can lay down the essentials of an adequate education for all men, everywhere, always. The scientific approach, on the other hand, is interested in discovering what the nature of man is, not in terms of an absolute essence, but *in terms of a developing career in time* and in relation to the world of things, culture, and history of which he is an inseparable part. It recognizes man's nature not as a premise from which to deduce the aims of education, but as a set of *conditions* which limit the range of *possible* educational aims in order to select the best or most *desirable* from among those for which man's nature provides a ground. An education should not be what it cannot be; it can be what it should not be; it may be what it should be.

In this chapter I shall briefly indicate an experimental approach to the question of educational ends and their relation to human nature. In the next, I shall consider the opposing claims made for a currently fashionable metaphysical view.

There are at least three distinguishable, but not separable, aspects of man's nature that are relevant to the formulation of valid educational ideals. (a) First, man is a biological organism subject to definite laws of growth. Certain powers and capacities mature, flourish, and decline according to a definite cycle. (b) Second, man is a member of society, heir to a cultural heritage and social organization that determine the forms in which his biological needs and impulses find expression. (c) Third, man as a personality or character exhibits a distinctive pattern of behavior, rooted in biological variation and influenced by the dominant norms of his culture, which he gradually develops through a series of successive choices.

Given these threefold aspects of man's powers, what ends of education should be stressed, and why? We say ends, rather than end, because an education that is relevant to at least these three aspects of human nature will have plural, even if related, ends.

(a) In relation to the development of the human organism, physical and mental, a desirable education takes as its end *growth*. By "growth" I mean the maturation of man's natural powers toward the highest desirable point which his body, his mind, and his culture make possible. It is a process which results physically in a state of health, and intellectually in a continuing activity of self-education.

The maturation of body and mind is natural; but so is stunting and retardation. Therefore, in selecting growth as an end, we are not *deducing* what should be from what is but are choosing the preferred consequences of one mode of action rather than another. There are many societies in which the development of certain features of the body and powers of the mind is not encouraged. For the same reason, since there are multiple possibilities of development, in selecting growth we are selecting a certain type or kind of development.

Growth, as everyone knows, has been emphasized by John Dewey as one of the central aims of education. But, as soon as one speaks of growth, critics who approach this end as if it were being urged in isolation from others are sure to inquire: growth in what direction? There is criminal growth, fascist growth, cancerous growth. From the fact that a thing is, it doesn't follow that it must or should grow. From the belief that a thing should grow, we do not yet know what direction the potentialities of growth should be encouraged to take. The necessity for a social frame of reference is clearly indicated as soon as we select growth as an educational end.

No one has seen this more clearly nor stressed it more insistently than John Dewey. From the very outset the end of personal growth has been allied with the social end of democracy in his educational philosophy. "This idea [that the object and reward of learning is continued capacity for growth] cannot be applied to *all* the members of a society except where intercourse of man with man is mutual, and except where there is adequate provision for the reconstruction of social habits and institutions by means of wide stimulation arising from equitably distributed interests. And this means a democratic society."[2]

Education for growth, then, goes hand in hand with education for democracy and a justification of one is tantamount to a justification of the other. But why continuous growth even if democracy is accepted as a social goal? There are at least two reasons. One flows from the nature of the democratic ideal, which is incompatible with fixed social divisions. It cannot function properly where individuals are trained independently of their maturing powers and possibilities of development. The second is that a world in which continuous growth is encouraged is more likely to make for the diversification and enrichment of experience than a world where individuals remain at the same level they have reached at the close of their schooling, learning nothing new even if they forget nothing old.

(b) We have already seen that every choice we make in selecting and fortifying certain tendencies among the plurality of potentialities in the individual must be undertaken from the standpoint of some social philosophy, or some ideal of social organization. What, then, are the grounds for our choice of the democratic social philosophy? Here, also, as in the case of the justification of ends, there are two generic approaches open to those who recognize the validity of the question—a metaphysical or religious "demonstration" ultimately based on absolute intuitions, and an empirical approach which regards the test of consequences as decisive.

The metaphysical and theological premises from which the validity of democracy has been allegedly derived are of the most heterogeneous variety. Many of them are mutually incompatible. They have been offered by polytheists, monotheists, atheists; Jews, Mohammedans, and Christians; Catholics, Lutherans, and Unitarians; and by philosophers of diverse schools. This suggests that the conviction with which the democratic ideal is held rests not so much on alleged metaphysical presuppositions that are beyond the test of experience, but on the actual or anticipated values of democracy in experience as contrasted with nondemocratic alternatives. It is interesting to observe that these *nondemocratic* alternatives historically have been justified by the identical metaphysical and theological presuppositions which have been advanced as the alleged premises on which democracy rests. And since these premises are compatible with social philosophies that are mutually contradictory, the latter cannot be derived from the former.

The existence of democratic communities in which individuals of conflicting religious faiths and metaphysical beliefs sincerely co-operate in democracy's support indicates that it is possible to find criteria for accepting democracy that do not depend on revelation or intuition. Indeed, to claim that democracy is uniquely entailed by only one set of theological or metaphysical intuitions, and that no one can sincerely or consistently be a democrat who does not embrace them, is not only logically false—it imperils the very existence of a democratic community. For the nonempirical character of these intuitions makes it impossible to find a workable method by which conflicts among them may be resolved and uncoerced agreements reached. In matters of faith, each sect regards itself as illumined and all others as blind.

The empirical method which regards democracy as an hypothesis, warranted by its consequences for weal and woe, holds out some promise of reaching agreement provided human beings can be induced to follow its lead in social affairs as in physical affairs. If we ask, then, why we should treat individuals of unequal talents and endowments as persons who are equally entitled to relevant consideration and care—the central idea underlying democratic institutions—we can point to consequences of the following *type:* it makes for greater tranquillity, justice, freedom, security, creative diversity, reasonableness, and less cruelty, insensitiveness, and intellectual intolerance than any other social system that has so far been devised or proposed.[3] There are more widespread commitments among men to these values, and a greater agreement on the method by which evidence is reached concerning whether or not they are present in any situation, than to any metaphysical or theological system which allegedly underlies them. Any one of these values has been or can be challenged in the course of experience. Its rejection or vindication depends on whether or not it furthers other values. There is no last resting point, nor is there a circle. We rest at each problem, until a new one arises.

This may be and has been contested by those who assert that there are ultimate values which are inarbitrable and that in the end only a radically ungrounded choice can be made when these ultimate values conflict. Existentialism is one of a variety of philosophical positions which stress the alleged fact of these ultimate values as a reason for denying or limiting the relevance of rational, scientific inquiry to problems of moral conflict.

Whether there are "ultimate" values for which we can offer no further justification or good reason is something which cannot be settled by fiat. It is a question of fact, not of definition. If there are such ultimate values, they may all be equally objective even if not universal. And if there are such ultimate values, it is clear from the complex chains of justification which are offered in defense of myriads of policies and decisions, that they are very few in number. An overwhelming number of value conflicts would be still arbitrable in the light of some shared ultimate or terminal value.

Further, when we analyze judgments of value in the problematic contexts in which they are made, we invariably find in the structure of the situation a reference to what is the case or might very well be the case that has a bearing upon the validity of the judgment. Taken out of context, out of a real situation of danger and choice, the answer to the question: Should I live or not? may seem ultimate or arbitrary in the sense that no further justification can be given. Examined in the actual, living context in which a genuine problem arises whether one should live or die, a thousand good or bad reasons may suggest themselves for doing one or the other. Theoretically, it is *possible* that those who differ in their judgments of value in any specific situation may agree about all the facts involved and all the consequences for themselves and others likely to follow from the envisaged alternatives. If and when this is the case, we may speak of the difference in value judgment as ultimate. So far, I have never found a situation of strong value conflict in which this *is* the case. Conflicts over values seem always associated with conflicting assessments of causes and consequences. As far as the justification of democracy is concerned against its communist or fascist critics, I have always found that the argument seems to depend directly or indirectly upon judgments of fact. This would seem to indicate that the conflict of values is here not ultimate but penultimate.

(c) On the level of character and personality, the aim of education should be the development of intelligence. Here we reach the key value in the sense that it is both an end and the means of testing the validity of all other ends—moral, social, and educational. How is it to be justified? Why should we educate for intelligence? Once again, the answers divide into those which reply in terms of the antecedent nature of man, and those which point to the consequences of intelligence in use. These consequences are many and desirable. Intelligence enables us to break the blind routines of habit when confronted by new difficulties, to discover alternatives when uninformed impulse would thrust us into action, to foresee what cannot be avoided and to control what can. Intelligence helps us to discern the means by which to instate possibilities; to reckon costs before they are brought home; to order our community, our household, and our own moral economy. All this and more, in addition to the joys of understanding.

Whether man is intelligent, and how intelligent, and what conditions his intelligence, are empirical questions on which considerable evidence has accumulated. One might, of course, ask: What must the nature of man be in order for him to become intelligent? And if anyone can derive from the answer more illumination than he had before, we can reply: Man must potentially have the nature of a rational creature in order to *become* intelligent. How little this tells us is apparent when we reflect that it is tautological, except possibly for cases of mutation, to assert that a thing possesses potentially the qualities and relations it actually exhibits in the course of its development. Potentialities may not all be realized but, in a certain sense, everything realized may be regarded as potential prior to the moment of its actualization. Men are and may become unintelligent, too. Unintelligence (or stupidity) is therefore also an antecedent potentiality. But since, potentially, man is both intelligent *and* unintelligent, what we select as the trait to encourage depends not merely on its potentiality but rather on its desirability. And desirability is an affair of fruits, not of origins.

Growth, democracy, and intelligence are related, inclusive aims which obviously embrace all but two of the educational ends enumerated at the beginning of this chapter. The place of vocational education will be considered in a separate chapter. Here I should like to say a few words in justification of the trait of independence. In every society, not excluding democracy, there is a certain suspicion of the nonconformist and a fear of appearing different. "Man," Santayana somewhere writes, "is a gregarious animal and much more so in his mind than in his body. He may like to go alone for a walk but he hates to stand alone in his opinion." This is not less true for our modern socialized world in which powerful pressures are making for uniformity of taste and opinion. The intellectual heretic, the protagonist of the variant in arts and letters, the unintegrated and unpoliticalized talent, run greater risks today of being victimized by hostile sanctions, official and unofficial, than in the last century when our economy was more loose-jointed. To strengthen intellectual courage—that rarest of virtues—so that the individual may more readily withstand the tyranny of fashions is to increase the variety and goods in experience. Despite social hostility to the intellectual outsider, his work not infrequently redounds to the benefit of society. Even where it does not, independence gives the individual a certain distance from his own work and prevents him from becoming *merely* a public or political character. There must be some private altars in a public world where the human spirit can refresh itself. A liberal education should enable individuals, without failing in their social responsibilities, to build such altars and to nurse their flames.

What I have been doing is illustrating the mode of procedure which an experimentalist follows in justifying the ends of education. Different ends may be proposed but intelligent decision among them can be made only by canvassing their consequences in experience. These ends may, of course, be supported by theologians and metaphysicians, too. They may *add* reasons drawn from their private store of principles to justify supporting them. But from the point of view of the experimentalist, it is illegitimate to make these supplementary reasons necessary articles of faith without which the ends in question cannot be consistently or sincerely held. In the light of the history of thought, it is clear that agreement on certain social and moral ends is possible among men whose theological and metaphysical presuppositions are incompatible with each other. It is also clear that it is possible for men to share the same set of presuppositions, and yet to invoke them in support of ends and practices of an antithetical character. This suggests that our grounds for accepting or rejecting human values are actually independent logically of our grounds for accepting or rejecting their alleged presuppositions. It further suggests that the evidence drawn from the fruits and consequences of the way in which ideals function in experience is far more warranted than the evidence for theological or metaphysical assumptions. The possibility is therewith established of broadening the area of moral and social agreement among men and building a better world on human foundations long before agreement has been won on first or last things.

NOTES

1. Adolf Hitler: *Mein Kampf* (English ed.; New York: Reynal and Hitchcock; 1949), p. 668.
2. John Dewey: *Democracy and Education* (New York: The Macmillan Company; 1916), p. 117.
3. For an amplification of this point, cf. my essay, "Naturalism and Democracy," in *Naturalism and the Human Spirit* (ed. Y. H. Krikorian; New York: Columbia University Press; 1944), pp. 40–64.

6. Educational Values and Goals: Some Dispositions to Be Fostered

William K. Frankena

I

There has been much impatience with what R. S. Peters calls "the endless talk about the aims of education," but this talk continues to go on, and we are invited to add to it on this happy occasion. Indeed, those who deny that education has ends or that educators must have aims seem always to end up talking about much the same thing in a slightly different idiom. At any rate, I am quite ready, at least on this occasion, to assume that there are values or goals which it is the business of education to promote, whether they are external, imposed, and far-off, or internal, autonomous, and nearby. I shall also assume that the values or goals which education is to promote consist of certain abilities, dispositions, habits, or traits. To have a single term for them I shall call them "dispositions," taking this word not in the narrower ordinary "sunny disposition" sense, but in the wider one common among philosophers.[1] So far as I am aware, there is really only one view that might reject the concept of dispositions in this sense, namely, existentialism,[2] and, as we shall see, even it seems to advocate our developing certain dispositions or, if you prefer, choosing certain postures.

Which dispositions is education to foster, then? Desirable ones, of course, since we do not call the formation of undesirable dispositions education or learning, but which ones are these? One answer is that the task of education is to foster the dispositions desired or regarded as desirable by the society doing (or paying for) the educating. This answer has a certain practical realism about it, but it is hardly a philosophical one, since it equates the

Reprinted, by permission, from *The Monist* 52 (January 1968): 1–10.

desirable with what is desired or thought to be desirable. A properly philosophical reply would say that the desirable dispositions are those required either for the moral life, the life of right action, or for the good life, a life of intrinsically worthwhile activities. I shall not try now, however, to present even an outline of the content of the good life, of the requirements of the moral life, or of the dispositions to be fostered in view of them. Instead, I shall take a somewhat different approach, also philosophical in a sense, which strikes me as interesting. I shall look at the three main movements in recent philosophy to see what dispositions they advocate our fostering. What interests me is that, different as they are, these three movements seem to offer us, not indeed the same, but *supplementary* lists of dispositions which we may combine, perhaps with certain corrections and additions.

In doing this I do not mean to aid and abet the tendency toward eclecticism already too far advanced in so-called philosophy of education, which I join Reginald Archambault and others in decrying.[3] I should hope and maintain that the promotion of the dispositions to be mentioned can be defended on a properly philosophical basis and must be eschewed if it cannot be. On the other hand, I think that it can be defended on more than one philosophical basis, and that we may agree on the following lists of dispositions even though we start from different philosophical premises. This is one reason why I am taking the approach I do. Each of us must have a philosophy of his own which serves as the basis of his thoughts on education, not just an eclectical synthesis of philosophical phrases and social science jargon, but, at least in the area of public education, formal or informal, we must manage some kind of agreement in our working conclusions about the dispositions to be promoted.

In what I shall say I can of course, only be rough and suggestive, rather than accurate or complete.

II

The three movements I refer to are (1) Deweyan experimentalism, instrumentalism, or pragmatism, (2) analytical philosophy, and (3) existentialism (*cum* phenomenology). These, apart from Thomism and Marxism, are the main currents in western philosophy today. Let us consider first Deweyanism, the philosophical movement most familiar to American educators. What is characteristic here, so far as education is concerned, is its emphasis on what Sidney Hook calls, "the centrality of method," the method of reflective thinking, scientific intelligence, or experimental inquiry. The concept of the method, with its five stages, is well known. It is thought of as *the* method of thinking, and the main task of education is regarded as that of fostering the habit of thinking in this way in all areas of thought and action. Get the power of thinking thus, Dewey virtually said, and all other things will be added unto you. What interests me now is the fact that this habit of thought is conceived of as involving a whole family of dispositions: curiosity, sensitivity to problems, observational perceptiveness, regard for empirical fact and verification, imaginative skill at thinking up hypotheses, persistence, flexibility, open-mindedness, acceptance of responsibility for consequences, and the like. Being associated with thinking, these dispositions have a strongly intellectual cast, though Deweyans reject the distinction between intellectual and moral dispositions and think of them as at least quasi-moral—and sometimes stress this practical aspect of them so much as to be charged with *anti*-intellectualism. They are dispositions whose matrix is the practice of empirical science. If we assume that this practice is one of the things human beings must be good at, then we may take this family of dispositions as among those to be fostered, even if we do not conceive of them exactly as Dewey did.

Analytical philosophy comes in various styles and must not be identified with either the logical positivism and therapeutic logicoanalysis of yesteryear or the ordinary language philosophy of today. In one style or another it has become more or less dominant in British and American philosophy, and is beginning to be influential in the philosophy of education.[4] Now, analytical philosophers of all sorts have tended to adjure the actual making or propounding of ethical, normative, or value judgments, and to be chary about laying down aims or principles for education and about making educational recommendations. They tend to limit philosophy to conceptual and linguistic analysis and methodological clarification. This attitude has been relaxing lately, but, in any case, there is a set of dispositions which are held dear by all analytical philosophers, no matter how purist: clarity, consistency, rigor of thought, concern for semantic meaningfulness, methodological awareness, consciousness of assumptions, and so on. These dispositions have been nicely characterized as "logical values" or "values in speaking [and thinking]" by J. N. Findlay.[5] Typically, analytical philosophers think, with some justice, that these values have been neglected both in theory and practice by Deweyans and existentialists, as well as by speculative philosophers, Hegelian, Whiteheadian, etc.—and especially by nonanalytical philosophers of education. Whether they are right or wrong in this, it does seem clear that their values should be among our goals of education at all levels. The title of a recent book proclaims that clarity is not enough,[6] and perhaps it is not, but it is nevertheless something desirable, and even imperative, both in our thinking about education and in our thinking about other things.

Existentialism is characteristically opposed both to analytical philosophy (though there are now some attempts at a rapprochement between these two movements) and to pragmatic empiricism. It is suspicious, among other things, of the "objectivity" so much prized, in different ways, by these other two movements. The implications of existentialism for education have begun to get attention from O. F. Bollnow and others,[7] but this is not what concerns me now. What interests me is that existentialism presents us with a third family of dispositions to be fostered: authenticity, decision, commitment, autonomy, individuality, fidelity, responsibility, etc. These are definitely moral (or, at any rate, "practical") dispositions as compared with the more intellectual, logical, or scientific ones stressed by Deweyan and analytical philosophers; but they are moral or practical dispositions that relate to the *manner* of life rather than to its *content*—not to *what* we do so much as to *how* we do it. To quote a recent writer: "Existential morality is notorious for its lack of content. But it does not cease for that reason to be morality. Everything is in the manner, as its sponsors would, and do, say."

As one of these sponsors does say:

> Value lies not so much in what we do as in how we exist and maintain ourselves in time . . . words like *authentic*, *genuine*, *real*, and *really* . . . express those more basic "existential values," as we may call them, which underlie all the valuable things that we do or say. Since they characterize our ways of existing in the world, they are universal in scope, and apply to every phase and region of our care. There is nothing that we say, or think, or do that may not be done either authentically . . . or unauthentically. . . . They are not "values" at all, in the traditional sense of this term, for they cannot be understood apart. . . . They are patterns of our lived existence in the world.[8]

I do not wish to suggest that an "existential" manner or posture is enough, and certainly not that we should be in a state of "anxiety" all the time, but it does seem plausible to maintain that there is a place in education for the development of such "existential" virtues along with others supplementing or even modifying them. There is at least *some* point in "the underground man's" remark in Dostoevsky's *Notes*, "perhaps, after all there is more 'life' in me than in you."

III

Thus we see that even though representatives of all three of our philosophical movements are typically reluctant to "talk about the aims of education," each movement itself enshrines or espouses certain dispositions that may well be included among the aims of education by those of us who do not mind such talk. The three philosophies are in general opposed to one another, and one cannot simply combine them, but the dispositions they value may be combined and included in our list of those to be cultivated in education, though perhaps not without some pulling and hauling. This is the main point I wish to make. I should like, however, to subjoin a few further points.

(1) Of course, we can espouse the Deweyan list of dispositions, even if we do not conceive them exactly as he does, only if we assume that empirical inquiry of a scientific kind is a good thing—sufficient, necessary, or at least helpful to the good or the moral life. This, however, is an assumption that would be denied only by certain extreme kinds of rationalism, irrationalism, and other-worldliness. In the same way, an adoption of the analytical philosopher's list of dispositions as among those to be cultivated involves assigning at least a considerable value to clarity, rigor, etc., an assignment which only an extreme irrationalist could refuse to make, though, of course, those who do make it will not all have the same conception of clarity or rigor. As for the existential virtues—it looks as if they can and must be accepted in some form by almost anyone who takes morality or religion seriously, that is, by anyone whose approach is not purely aesthetic, conventional, legalistic, or spectatorial.

(2) It seems to me that existentialism and its sisters and its cousins and its aunts do not put sufficient store on rationality, meaning by this roughly the set of dispositions prized by the Deweyans and the analytical philosophers taken together. Indeed, they tend to suspect and impugn it. Yet, even if we confine ourselves to the *how* of our approach to life and let the *what* take care of itself, it seems at least irrational to neglect the virtues of logic and science. To quote Israel Scheffler, "We are . . . faced by important challenges from within and without . . . whatever we do, we ought, I believe, to keep uppermost the ideal of rationality, and its emphasis on the critical, questioning, responsible, free mind."[9]

(3) Even so, the existentialists (and their sisters and their cousins and their aunts) are perhaps right in feeling that the values of rationality must at least be supplemented by those of commitment and engagement, as S. T. Kimball and J. E. McClellan have argued,[10] along with many others who think that our western culture is in danger of being overcome by its "committed" opponents. (In this perspective it is a bit ironical that our "uncommitted" are precisely those who are most attracted by existentialism.)

(4) Of course, if we try to combine rationality and commitment—the first without the second being empty and the second without the first blind—we must find some teachable kind of union of open-mindedness and belief, of objectivity and decision. This is one of the crucial problems of our culture, as has often been pointed out.

(5) In my opinion, none of the three families of dispositions includes enough emphasis on sheer (not "mere") knowledge, the intellectal virtue so esteemed by Aristotelians—not just knowing *how* (which was given a big boost by Gilbert Ryle) but knowing *that*, the kind of knowledge contained in the findings of history, science, and other cognitive studies (including knowing *why*). One reads, for example, that a college education must have as its goals "intellectual initiative and mature self-reliance," a statement which roughly synthesizes Dewey and existentialism, but there is enough "formalism" in me to make me convinced that education ought to promote not only certain "qualities of mind [and character]," but also certain "forms of knowledge," even if the knowledge must sometimes be second-hand and not acquired "by doing."[11] To parody Bertrand Russell, the good life, moral and otherwise, is a life inspired by certain qualities of mind and character and guided by knowledge, actually possessed knowledge *that* is important both for the guidance of action and for the content of the good life. I therefore feel some agreement with Maritain when he writes that contemporary education has too much substituted "training-value for knowledge-value . . . mental gymnastics for truth, and being in fine fettle, for wisdom."[12] As Jerome Bruner puts it: "Surely, knowledge of the natural world, knowledge of the human condition, knowledge of the nature and dynamics of society, knowledge of the past so that it may be used in experiencing the present and aspiring to the future—all of these . . . are essential to an educated man. To these may be added another: knowledge of the products of our artistic heritage. . . . "[13] John Stuart Mill was, no doubt, right in attacking education that is "all *cram*" and does not provide "exercises to form the thinking faculty itself," but he went too far in adding, " . . . the end of education is not to *teach*, but to fit the mind for learning from its own consciousness and observations. . . . Let all *cram* be ruthlessly discarded." For Mill himself goes on to insist that each person must be "made to feel that . . . in the line of his peculiar duty, and in the line of the duties common to all men, it is his business to *know*."[14] It seems to me to follow that there is place in education for some "teaching" and even some "cram." I grant it may be that, if we seek first to form the thinking faculty itself (i.e., certain qualities of mind), then all other things will eventually be added unto us, including knowledge. But must we wait until after school is over for them to be added? *Can* we?

A recent cartoon about education has a father saying to his child sitting in a high chair with his food before him, "Think. Assimilate. Evaluate. Grow." It seems to me that this is to the point as a spoof of a certain conception of education, since the word "Grow" does not add anything, and that a more sensible view would say, "Think. Assimilate. Evaluate. Know."

(6) There are, of course, certain other sorts of dispositions that must be added to the three families indicated above as goals of education. There are,

first, moral dispositions relating to *what* we do and not merely to *how* we do it, e.g., benevolence and justice (i.e., knowing *what* to do and being disposed to do it), second, the dispositions involved in aesthetic appreciation, creation, and judgment (not just "knowledge of the products of our artistic heritage"), and third, the dispositions required by the democratic way of life, so far as these are not already covered.

(7) In what I have said thus far, I have had *public* education primarily in mind. *Private* education, formal and informal, may add still another group of dispositions, namely, those involved in religious faith, hope, love, and worship. However, some care is needed, perhaps even some reconstruction, if one proposes to combine a Deweyan emphasis on scientific intelligence or an analytical philosopher's emphasis on clarity and rigor with anything like a traditional theistic faith. If one proposes to foster such a faith, one must at least give up trying to cultivate also a disposition to rely on logic and science *alone* as a basis of belief and action. If one wishes to insist on the necessity of the latter disposition, one must reconstruct the traditional conception of religion and God—as Dewey did in *A Common Faith*. Of course, if one means by religion merely some kind of basic commitment or other, or any kind of ultimate belief about the world whatsoever, or a vague "duty and reverence" (as Whitehead does),[15] or simply whatever an individual does with his solitude (Whitehead again), then all education is and must be religious (as Whitehead says), even an atheistic or militantly antireligious one. Then "the Galilean" has indeed conquered, but then he has also become very, very pale—so pale as to be indistinguishable from or to his opponents. For, even if one has a "religious" belief in this wide sense, one may still also believe that

> . . . beyond the extreme sea-wall, and
> between the remote sea-gates,
> Waste water washes, and tall ships flounder,
> and deep death waits. . . .

as Swinburne did.[16] I say this because of what one finds in some discussions of the place of religion in public schools, where, from the premise that every ultimate belief is a religion, the conclusions are drawn, first, that religion both is and should be taught in public schools, and, second, that therefore theism (or Catholicism, Protestantism, and Judaism) may and should be taught there. As for private schools and colleges—whether they should in fact foster religious dispositions in a narrower theistic sense is too large a question to treat here; the answer, I suppose, assuming that there should be private schools and colleges at all, is that it depends on the purposes for which they exist.

NOTES

1. Dewey prefers the word "habit" but uses "disposition" also. See *Human Nature and Conduct* (New York: Henry Holt, 1922), end of Pt. I, Section I: *Democracy and Education* (New York: Macmillan paperback ed., 1966), p. 328.
2. See O. F. Bollnow, *Existenzphilosophie und Pädagogik* (Stuttgart: W. Kohlhammer Verlag, 1962), pp. 14ff.
3. R. Archambault (ed.), *Philosophical Analysis and Education* (London: Routledge and Kegan Paul, 1965), p. 3.
4. See *ibid.*, pp. 6, 13.
5. See *Language, Mind and Value* (London, 1963), pp. 105–127.
6. H. D. Lewis (ed.), *Clarity is Not Enough* (London, 1963).
7. See Bollnow, *op. cit.;* G. F. Kneller, *Existentialism in Education* (New York, 1958); Van Cleve Morris, *Existentialism in Education* (New York, 1965).
8. See respectively, C. Smith, *Contemporary French Philosophy* (London, 1964), p. 229; J. D. Wild, *Existence and the World of Freedom* (Englewood Cliffs, N.J.: Prentice-Hall, 1963), pp. 161–165. See also Whitehead's remarks on "style" in *Aims of Education* (New York: New American Library, Mentor Books, 1949), p. 24.
9. "Concepts of Education: Some Philosophical Reflections on the Current Scene," E. Landy and P. A. Berry (eds.), in *Guidance in American Education* (Cambridge, Mass.: Harvard University Press, 1964), p. 26.
10. *Education and the New America* (New York: Random House, 1962).
11. See P. H. Hirst in Archambault, *op cit.*, pp. 117f., for a discussion of this point.
12. *Education at the Crossroads* (New Haven, Conn.: Yale University Press paperback ed., 1943), p. 55; cf. Aquinas' remark that "The first thing that is required of an active man is that he know."
13. *On Knowing* (New York: Atheneum, 1965), p. 122.
14. See his essay "On Genius" in K. Price, *Education and Philosophical Thought* (Boston: Allyn and Bacon, 1962), pp. 455f.
15. *Aims of Education*, p. 26.
16. "Hymn to Proserpine." Or as B. Russell did in "A Free Man's Worship."

Part Three: Social and Political Values

Introduction

In the general Introduction, we noted that freedom, equality, and other political values are frequently utilized to justify social and educational policy. In addition, these values are, of course, presented to children as ideals worthy of adoption and are therefore important elements of moral education as well. Yet their significance for educational theory has often been taken for granted rather than systematically analyzed. Not so for R. S. Peters, however, whose discussion of freedom in *Ethics and Education* is related to a number of educational issues.

In the excerpts to follow, Professor Peters begins by pointing out that although freedom need not be accepted as an absolute value, those who would limit it are obligated to provide valid reasons for their interference. But why should there be "a presumption in favor of allowing people to do what they want"? Peters's reply to that question is a significant contribution to meta-ethical theory. "If a person is asking seriously what he ought to do or what there are reasons for doing," Peters argues, "he must obviously demand absence of interference in doing whatever there are reasons for doing." Or to put it another way, it would be unreasonable, or even irrational, not to insist on the freedom necessary to carry out the chosen course of action. Peters thinks that other fundamental moral values, including equality, respect for persons, and the consideration of interests, can be justified in the same manner. In other words, the final appeal in ethics is to reason rather than to fact, intuition, or emotion.

Other excerpts from Peters's analysis of freedom in this chapter include his comments on the "paradox of freedom" ("too much freedom leads to too little"), a discussion of the freedom of the child in relation to the duty of the teacher, and some reflections on the freedom of the parent with regard to private education. One noteworthy feature with respect to the latter is Peters's discussion of the resolution of conflicting values at the end of the chapter.

In Chapter 8, Charles Frankel begins by observing that political philosophers have often turned to the notion of equality of opportunity to extricate themselves from difficulties encountered in attempting to explain the principle of equality itself. The notion here, as Frankel observes, is that the claim that people are equal may be taken to mean that they should have equal opportunity to satisfy their wants. But linking equality to the opportunity to satisfy wants will work, says Frankel, only if wants are subjected to certain limitations. These limitations raise moral issues as well as practical questions, all of which Frankel considers at some length.

Of special interest to both philosophers and educators is Professor Frankel's analysis of the "meritocratic" conception of equality of opportunity (which he relates to "performance"), and the "educational" conception (which is associated with "the development of individual potentialities"). Having considered (1) various guidelines—including cost in the broadest sense, moral considerations, and "standards by which we measure either performance or ability"—and (2) the probable consequences of adopting either ideal, Frankel concludes that the educational conception is preferable to the other. His discussion of culturally biased tests in connection with standards of measurement speaks directly to the concerns of educational policy makers.

Professor Scheffler's essay is included here rather than in Part Five because his analysis of democracy and its implications for education is a natural extension of the discussions of freedom and equality in the first two chapters of this section.

Undergirding democracy as a social and political system, as Professor Scheffler notes, is the commitment to government by consent. Consent in turn implies informed choice freely given, and informed choice implies an educated populace.

Professor Scheffler contrasts the democratic ideal of an open, dynamic society, one that is receptive to change based on the reasoned deliberation of informed citizens, with the more historically familiar authoritarian model, in which there are sharp divisions between rulers and ruled, with the former bearing responsibility for establishing policy that the latter are expected to accept without question. With regard to both formal schooling and the more informal education provided by other social institutions, the contrast is between fostering critical thinking and inculcating doctrines decided upon in advance; between consent based on information and inquiry, and consent derived from indoctrination and intellectual manipulation.

Scheffler draws an instructive parallel between ethics and science in this essay. Moral judgments as well as scientific doctrines must be consistent with the best available evidence and supported with the best available reasons. Similarly, a commitment to the moral point of view entails a willingness to revise moral attitudes in response to new information or changing conditions, just as a commitment to the scientific method entails a willingness to revise theory in light of further inquiry. As Scheffler points out, the common denominator here is reasonableness. He writes, "It is evident, moreover, that there is a close connection between the general concept of *reasonableness*, underlying the moral and the scientific points of view, and the democratic ideal." It follows then for Scheffler that reasonableness ought to be one of the fundamental traits encouraged and cultivated by educators.

Professor Scheffler acknowledges that "such a direction in schooling is fraught with risk, for it means entrusting our current conceptions to the judgments of our pupils." It is, however, a necessary risk, for unless we are willing to subject our social, political, and moral beliefs to the independent judgment of our students, it is difficult to see how we can do more than merely pay lip service to the democratic ideal.

7. Freedom

R. S. Peters

* * *

1. JUSTIFICATION OF THE PRINCIPLE OF LIBERTY

The fundamental arguments for putting the presumption in favour of freedom derive surely from the situation of practical reason. If a person is asking seriously what he ought to do or what there are reasons for doing he must obviously demand absence of interference on doing whatever there are reasons for doing. Reasons for interference may, of course, derive from other principles; but other things being equal he must demand to be allowed to do what there are reasons for doing. Otherwise his deliberation about alternatives would have no point. Now such deliberation is not something that grows out of his head like a plant from a bulb. It mirrors a social situation into which he has been initiated in which alternative courses of action are suggested and discussed. In such deliberation assessments such as 'wise' and 'foolish' are applied to suggestions in the light of public criteria which are built into the form of discourse.

Given, then, the public character of the situation out of which practical reason develops and which gives meaning to the terms in which it is assessed, it would be strange and paradoxical for a person, who has taken this sort of discussion into his own mind, to hive it off permanently from the public deliberation from which it derives its very existence and meaning. Purely on grounds of prudence, too, if a person is genuinely concerned about what he ought to do, he would be very foolish to shut himself off from other rational beings who also have views about what there are reasons for doing. It would be even more foolish to impose constraints on others so as to prevent them from giving him advice. Spinoza argued that there is nothing more useful to man than other rational men;[1] for conversation with

Reprinted with permission of the publisher from *Ethics and Education*, 2d ed. (London: George Allen & Unwin LTD, 1970), pp. 180–182, 186–188, 192–198, 204–207.

other men is the most obvious means the individual has of increasing his understanding of the universe. The same point applies to the enhancement of an individual's judgment about the alternatives which present themselves in the practical sphere. The conclusion of this argument must surely be that freedom of expression—at least of other rational beings—must be demanded by any rational being who is genuinely concerned with answering the question 'What ought I to do?' as well as possible. For on a matter such as this, which is a matter not of private taste but of public criteria, he would be very stupid if he deprived himself of access to considerations which others might offer.

The argument, however, need not be based simply on the manifest interest of anyone who seriously asks the question 'What ought I to do?' For the principle of liberty, at least in the sphere of opinion, is also surely a general presupposition of this form of discourse into which any rational being is initiated when he laboriously learns to reason. In matters where reason is paramount it is argument rather than force or inner illumination that is decisive. The conditions of argument include letting any rational being contribute to a public discussion. For, as Mill pointed out long ago, truth must be the sufferer if opinions are stifled.[2]

So far the presumption in favour of the principle of freedom has only been established in the sphere of opinions. The presumption in favour of the principle has also to be justified in the sphere of actions. This is not very difficult to do; for the opinions expressed in practical discourse are intimately connected with actions. They are attempts to answer the question 'What ought I to do?' As has already been pointed out anyone asking this question must demand freedom of action for himself to do what there are reasons for doing; otherwise his deliberation would be pointless, a rehearsal without a play to follow. But must he also demand such freedom for others?

The same sort of considerations surely apply here as in the justification in the previous chapter of the principle that one ought to consider the interests of others. If a person joins with other rational beings in trying to answer questions of practical policy and if, as a rational being, he must demand freedom of action for himself to do what there are reasons for doing, how can he engage in such discussions with other rational beings and yet deny to them what he rationally must demand for himself? If they are beings like himself who are rational enough to be concerned with the question 'What are there reasons for doing?' and if, as rational beings, they must make a prima facie objection to being interfered with in carrying out what there are reasons for doing, how can anyone treat others as rational beings in so far as their contributions to a discussion may be extremely pertinent and yet, at the same time, refuse to treat them as rational beings by interfering with their freedom without good reason? Because, therefore, in the sphere of practical reason, there is such a close link between discussion and action, in the sense that a rational man who asks the question 'What ought I to do?' must demand for himself freedom to do what there are reasons for doing, freedom of action as well as freedom of thought can be shown to be a general presupposition of practical discourse in so far as it is a public activity to which rational beings contribute.

* * *

3. THE PARADOX OF FREEDOM

It is one thing to agitate against the tyranny of arbitrary men or of public opinion; it is quite another to take practical steps to be rid of the constraints which they impose. For here we are confronted with what Popper has called the 'paradox of freedom',[3] which is that too much freedom leads to too little. In realms which are not either those of indifference or those where interference is almost impossible, if people are allowed to do what they like, what tends to happen is that the strong impose arbitrary constraints on the weak. In such spheres individuals are only in fact free to do or say what they like if they are protected from arbitrary interference by law or public opinion or both. The unpalatable lesson of history is that it takes a constraint to catch a constraint.

There are many who have yearned for a life where men would be able to do what they want without rules and regulations, without people in authority to give them orders. But such a state of nature is not a state where there are no constraints; it is a state where the constraints are arbitrary. Such constraints do not vanish into thin air by men being transformed into angels; they gradually cease if another less obnoxious and more levelling type of constraint is developed to protect the weak. No doubt many men would be decent enough, under favourable conditions, not to interfere with others. But laws are not made primarily to restrain those who follow the moral law within; they are made to protect ordinary people against those who acknowledge no such code or who abide by it only haltingly. 'For liberty', as Locke put it, 'is to be free from restraint and violence from others; which cannot be where there is no law.'[4] This is the theme of so many stories of the brief period when the West was being opened up, when the land was overrun by arbitrary adventurers. In such a state of nature the weak were oppressed by the strong until the rule of law was established. Similarly in the economic sphere unfettered free enterprise was fine for the few; but it was only when the countervailing constraints of the trade unions on the few became effective and when laws were introduced governing wages and conditions of employment, that the economically weak ceased to be grossly exploited by the strong. In the sphere of freedom of speech, too, we can only speak our mind in public because the law, backed by public opinion, protects us.

The individual, therefore, in order to have his freedom in some sphere guaranteed, has to accept a less obnoxious, more levelling, sort of constraint which protects him and which also constrains him from interfering with others in a similar way. The example of voting illustrates this very well. The act is simple—just the recording of a cross on a piece of paper. But a most elaborate system of electoral law has been found necessary to make sure that the voter is in fact free to record his vote as he pleases. In the early days of full adult suffrage all kinds of pressures were put on people to vote as someone else required. The system of electoral law was devised to protect the individual, to ensure his freedom in a field which is of vital importance for parliamentary democracy.

Some political philosophers have grasped clearly this 'paradox of freedom' and have drawn quite mistaken inferences from it; for they have argued that it demonstrates that 'freedom' means not doing what one wills without restraint but accepting the law or the 'real will' of the community. This is to confuse what looks like a general empirical condition of 'freedom' having concrete application with the meaning of 'freedom'. It may well be the case that, men being what they are, they will not in fact be free to do what they please in a certain sphere unless they are prepared to accept constraints prohibiting others from interfering with them, and them with others, in this sphere. But this does not show that 'freedom' *means* the acceptance of constraint. To take a parallel: it may be the case that men cannot think without brains; but this does not imply that 'having a brain' is part of the meaning of 'thinking'. Those who have confused such a general empirical condition of freedom with its meaning have often been led to do so because they have also been advocates of 'positive freedom', in which the passions of the individual are thought of as having been mastered so that his 'will' is free from internal constraints. As many such thinkers (e.g. Rousseau) have equated the 'real will' of the individual with the laws of the community working through him, these two strands of thought have become intertwined and the doctrine has emerged that the true freedom of the individual consists in obeying the laws of the state and that the individual may have 'to be forced to be free'.

Those who live in the modern world would need little reminding that the law, which is the most far-reaching and effective form of constraint yet devised by man, can oppress people as well as safeguard their liberties. The state can become a Leviathan as depicted by Hobbes or an instrument of an oppressing class, as depicted by Marx. It is a strange irony, however, that some political philosophers, who have favoured this intrusion of the state into the private life of the individual, have argued for such a condition by claiming that 'real freedom' demands it.

* * *

5. FREEDOM IN EDUCATION

So far an analysis has been given of the principle of liberty, an attempt has been made to justify it, and general problems of its application have been considered. Its application in the specific field of education must now be examined. The issues raised here are very different from those raised in the previous two chapters; for whereas they were concerned with the distribution and matter of education, questions to do with its manner have now to be tackled. There is, of course, an issue to do with the freedom of the parent in relation to educational provision, which is closely connected with the problem of its distribution; but the more perplexing problems relate to the freedom of the child and of the teacher. The problems are created by the differences in application necessitated by the fact that the principle has to be applied in an educational situation. We shall begin, therefore, with the freedom of the child, pass on to the freedom of the teacher, and end up with a few brief remarks about the freedom of the parent.

(A) THE FREEDOM OF THE CHILD

It has been maintained that the concept of freedom joins together two main components: first the notion of possible wants or decisions and secondly that of the absence of constraints upon them. In applying this concept to a situation in which adults are involved, their possible wants are taken more or less for granted as brute facts about them. A question of freedom arises only when there is a possibility of their being constrained in respect of such wants.

In an educational situation, however, the application of the principle is not so straightforward. For, almost by definition, it is a situation in which constraints are imposed upon children's wants. To start with children are compelled to attend school which is not a promising start from the point of view of freedom. Some of them, of course, want to attend as well; but the combination of legal requirement and parental pressure do not readily intimate an area of free choice and action.

Secondly the conditions under which learning takes place make it imperative that something like the rule of law should be established within an educational situation. However much encouragement is given to children within a class-room to follow their own interests and to work at their own pace, there must be conditions of order sufficient to permit a large number of children to work at the same time in a small space. What such conditions are depends on the age and number of children, what is being learnt, and the amount of space available. But without minimum conditions of order a class-room would degenerate into a Tower of Babel, and the freedom of some would be exercised at the expense of others. Education could not go on.

Thirdly an educational situation involves essentially a contrived and controlled environment. Children may be encouraged to choose things and to follow their own interests in their study. Indeed emphasis may be placed on the desirability of spontaneity and autonomy. But such choice has always to be exercised within a range of what is thought desirable. Amongst the materials, with which they are encouraged to experiment, knives and liquid paraffin are not included. Horror comics and pornography do not adorn the shelves of school libraries. Sexual experimentation is not encouraged in the common rooms. Such controlled conditions act as general constraints on the wants of children. By means of them adults exert a steady, if mild sort of pressure.

This must be the case; for no educator can take the wants of children for granted, part of his business as an educator being the transformation of wants, both in respect of their quality and in respect of their stability. To effect such a transformation a background of constraints on children's wants is necessary. It may, of course, be possible within such a controlled environment to harness the worth-while content of education to existing wants either by developing understanding, skill, and a sense of standards in relation to existing wants as in activities like canoe building, making guitars, or dancing, or by using such existing wants as a means to developing new ones, as in the project method. But it may at times be necessary to exert pressure on children so that they master something irrespective of what they want. Many students, for instance, have been told to write an essay on something in which they were not at all interested and, as a result, have developed a new interest in something. Such possibilities open up wider empirical questions about intrinsic and extrinsic motivation as well as moral questions about the permissibility of various methods of changing people's wants. But the basic point remains: no educator can be indifferent to what children want. He cannot, as in an ordinary social situation, say that what people want to do is their own affair, provided that they do no damage to others or interfere with their liberty. To adopt this *laissez-faire* attitude in a school would be to abdicate as an educator. Caretakers, maybe, can adopt such an attitude, but not teachers.

The justification of such a framework of order does not fall, naturally enough, in the main under the principle of liberty. It falls under the principle of the promotion of what is good, with its subordinate principle of the consideration of what is in people's interests. Nevertheless a case can be made in terms of liberty for such conditions of order in school which illustrates the paradox of freedom very well. If the rule of law imposed impartially by those in authority is absent it simply is not the case that children are actually able to do what they individually want. They are subject either to the arbitrary will of a bully or to the tyranny of peer-group pressure. Progressive schools in which the staff, as a matter of policy, withdraw from their proper function of exercising a just and levelling form of social control, are notorious for peer-group pressure and the proliferation of rules administered with severity by the children themselves. Or something like a state of nature prevails, as depicted with ghoulish exaggeration by William Golding in his *Lord of the Flies*. I once asked a colleague why his parents took him away from one of these schools. He replied that it was such hell when the headmaster was not around because of the bullying that went on. The headmaster in question prided himself on the fact that the staff laid down no rules. What he had not grasped was that there are other ways in which adults can control situations, when they are there, without having to formulate rules.

In all such situations, when human beings are gathered together, it is completely unrealistic to suppose that men or children are ever, as a matter of fact, free to do what they like simply because of the inherent decency and good sense of all concerned. They are in fact free because some rule or other, in addition to the moral law, is effectively enforced which prohibits interference. The practical choice is never between simply doing as one likes and being constrained; it is rather between being subject to different types of constraint. From the point of view of freedom it is a better bet for the individual to accept a system of levelling constraints which limit his freedom of action but limit also the freedom of action of others to interfere with him, than to commit himself to a state of nature in which he runs the risk of being arbitrarily coerced or subjected to merciless group pressure. Adolescents who join in rejecting the authority of parents and teachers often make a dubious gain in terms of freedom; for they find that their opportunities for doing what they want as individuals are much more stringently curtailed by peer-group pressures.

The gist of the argument so far has been to show that because an educational situation is one in which an environment is specifically contrived so that what is good can be promoted and passed on, there are conditions of order which must obtain in order that this overriding aim can be implemented. As, however, freedom is an independent principle like justice it cannot be abrogated entirely for the sake of the promotion of what is good. The presumption in favour of it still holds. It has to be shown that restrictions imposed on children, even in this contrived situation, are essential in terms of the promotion of what is good. Rules for the sake of rules, or as expressions of a teacher's love of power, are anathema to any rational man. Some types of rules, it has been argued, can be justified in relation to the paradox of freedom, as well as because of their necessity for the promotion of what is good. But the fewer restrictions the better, and all such rules must have some point. What they are, how they are decided upon, and how they are enforced will depend upon all sorts of contingent questions to do with the age of the children, the size of the school, whether or not it approximates, as does a boarding school, to a 'total institution', and so on. Contingencies abound when such a concrete level of the implementation of general principles is reached. About such contingencies there is very little, philosophically speaking, to say; for common sense must not be passed off as philosophical analysis. Philosophy is only one component in educational theory. When it comes to the formulation of particular principles for implementation in particular schools the contribution of philosophy, in terms of abstract analysis and justification of principles, must be put together with that of psychologists, sociologists, and teachers, who have practical experiences of the particularities in question. Philosophy contributes to practical wisdom but is not a substitute for it.

There is, however, another aspect of the freedom of the child which is very relevant to the ideals of progressive schools; for often such schools are not much concerned with providing conditions in which education of what they derisively call an 'academic' sort can proceed. They are much more concerned with the moral rehabilitation of misfits, or with character development, and emotional maturity. They are particularly interested, perhaps, in the development of moral autonomy and independence. Children have to learn to make their own choices, to stand on their own feet; it is argued that this sort of permissive atmosphere encourages them to do this.

Anyone concerned seriously with answering the question 'What ought I to do?' will be equally concerned about implementing such a laudable ideal; for the search for reasons for action is the hall-mark of the autonomous person. He is not prepared to accept authoritative pronouncements and is unhappy about simply doing what others do without inquiring any further. He will thus be very concerned that people should be educated in a way which encourages independence of mind. But he will also hope that this can be done without too great a cost in terms of security and happiness—especially in view of the danger of a relapse into submission to authority or group conformity, because of the 'fear of freedom' already mentioned. He will be concerned that individuals develop who can make proper use of the areas of discretion permitted by a liberal society. Little, however, is yet known about the conditions which favour the development of such autonomy. Do children in fact learn to behave autonomously by being brought up from their earliest years in a very permissive atmosphere without a proper framework of order?

This seems highly improbable both on general grounds and on the basis of the slender empirical evidence that there is about such matters. It would not be appropriate to enter into a detailed consideration of the empirical evidence, but the general grounds can be briefly mentioned as they make explicit considerations which are often forgotten in talk about children's autonomy and choice.

Autonomy implies the ability and determination to regulate one's life by rules which one has accepted for oneself—presumably because the reasons for them are both apparent and convincing. Piaget has shown that such an attitude towards rules is generally impossible before the age of about seven. Secondly there is evidence to suggest that the giving of reasons for action has little educative effect at an early stage. Yet long before this age children, for reasons both of their own survival, and for the welfare of others, must acquire in *other ways* a basic code of behaviour.[5] Thirdly they have to learn what it is to act on rules generally before the notion of determining their own code of rules for themselves can have any significance for them. If, too, such a choice is to be a real possibility for them rather than a lip-service to a shibboleth, they must have experience on the basis of which they can decide between alternatives. This is parallel to the problem of an 'elective curriculum'. What sense is there in saying that children should have 'choice' of what subjects they are to study unless they have been given some experience of such subjects on the basis of which they are in a position really to 'choose'? There is all the difference in the world between choosing between alternatives and 'opting' for alternatives on the basis of what is immediately attractive. I was once discussing with a girl her 'choice' of a College of Education. It transpired that she was inclined towards one college rather than another because of the picture of it on the brochure! Choice should not be confused with the way of opting for things which is encouraged by advertising agencies. For children to learn to choose in this real sense they must live in a fairly predictable environment so that they can learn to make realistic assessments of consequences. In human affairs the environment is mainly social, i.e. constituted by rules and standards. If predictability is not provided by something approximating to a rule of law how can the capacity for choice be developed in children? If rebellion and criticism are regarded as valuable something concrete must be provided to rebel against, so that children can learn to do this, knowing what is likely to happen. The virtue of a legal system, in relation to its encouragement of autonomy, is that it provides rules with determinate sanctions. Individuals who break the law do so with the knowledge of what is likely to happen if they are found out. They are thus put in a position where concrete probabilities are presented to them. If children are brought up without such stable conditions it is difficult to see how they can learn to make realistic choices.

Fourthly those who are staunch advocates of autonomy, such as Existentialists, draw attention both to the necessity and virtue of the choice between rules which conflict on particular occasions. It is seldom pointed out that such conflicts are only real if the individual already accepts the rules which conflict as binding on him. But if he has not been brought up in such a way that he has internalized a set of rules, how could such conflicts ever occur out of which his independence and strength of character begin to develop?

Fifthly although there is little virtue in order for its own sake, the fact is that human beings are beset by anxiety unless minimum conditions of order are provided. If they are anxious or insecure they will be an easy prey to all sorts of irrational pressures. They will submit readily to authoritarian regulation or peer-group pressure; they will dart to and fro between alternatives that offer easy gratification. They are in the worst possible condition to learn how to direct their lives autonomously.

The development of autonomy is a slow and laborious business. Young people have to learn gradually to stand on their feet and direct their own lives. They are not likely to do this if they are not encouraged to take responsibility and make choices about important matters within their limited experience; still less are they likely to do it if they are pitchforked into an anarchic situation in which they are told that they have to decide everything for themselves. Rationality requires a middle course between authoritarianism and permissiveness. Above all what is required is a rational attitude to and exercise of authority on the part of parents and teachers. Further consideration of this must be reserved until Chapters IX and XI.

* * *

(C) THE FREEDOM OF THE PARENT

The ground of a parent's right to educate his children as he thinks fit is not altogether perspicuous. It presupposes all sorts of things about the parent-child relationship which would take a long time to discuss. Some regard the parent's rights in this sphere rather like rights to property; others assimilate them to those of a trustee. But though the grounds of this right are obscure there are few in Britain who would question that the parent has such a right. Indeed in England the right of a parent to educate his child as he likes is one of the most long-standing rights of the subject. Burke, for all his attack on 'abstract rights', includes the right to 'the nourishment and improvement of their offspring' amongst the 'real rights' of the subject. Even Hobbes, thought to be one of the precursors of modern totalitarianism, presumed that the sovereign would not interfere with matters such as the subject's choice of vocation, choice of wife, and choice of education for his children. This right is of particular importance for those whose religion requires a special sort of education.

This right was established, of course, long before compulsory education for all was ever contemplated. Such compulsion was obviously enough an infringement of liberty in so far as parents were no longer free to exercise their discretion by not having their children educated at all. The arguments for such a limitation of freedom are obvious enough too. A highly industrialized society cannot be run effectively unless a certain level of literacy and numeracy is attained by all and a high degree of specialized knowledge and skill by many. Also without certain minimum conditions being insisted on, individuals would be deprived of the necessary conditions for attaining the quality of life in such a society which is handed on in a good educational system. Compulsion cannot force people to become educated. But without it many children would be deprived of their opportunities for entering into their cultural heritage. Such a limitation on freedom can therefore be justified both in terms of the interests of the community and the long-term interest of individuals potentially composing it. For matters as important as this cannot be left purely to the good sense of parents.

Since the development of state education the right of the parent has come to mean that of educating his children in a private school if he should wish to spend his money in this way. For most people such a right is 'formal' rather than 'actual'; for only a small minority have the money necessary to exercise it. Naturally enough the state cannot be expected to pay for such a luxury and naturally it is reasonable for the state to insist that a certain minimum standard of educational provision should be maintained. In this way the point of a compulsory system is retained without insisting on uniformity in implementation. For without this latter condition the children themselves might suffer as a result of the parent's exercise of their freedom of choice and so, derivatively, would the state.

There are few who would dare challenge this watered-down prima facie right of parents in a democratic society. Arguments against the existence of private schools derive from a conviction that it is unfair that the possession of money should put some in a privileged position, especially as much of the wealth so used is inherited rather than earned by the parents in question. When so few have actual access to a formal right of this sort it is argued that there is a case for abolishing it. Fairness must take precedence over freedom—especially as the present division in the educational system tends to perpetuate class attitudes which encourage lack of respect for persons.

At this point another principle becomes relevant—that of the promotion of what is good; for one of the main reasons why parents, who can afford it, insist on exercising their right is that the quality of education provided in the best independent schools is very good indeed. Of course there are many exceptions. Some independent schools, especially at the preparatory level, are not very good, and some parents have predominantly social motives for exercising their right of choice at such a great expense. They do not want their children to mix with children from another social class; or they see that education at an independent school may enhance their children's prospects of high social status and good employment. But the fact is that there are many very good independent schools with long traditions which have built up by long experience and trial and error a system which successfully combines academic achievement with character-training. Historically these schools have been pioneers in many spheres of educational experiment and the state system owes an enormous amount to their example. It would be criminal folly to dismantle a system of education which contributes and has contributed so much to the quality of education in England. The real scandal is that both the major political parties and the public as a whole have cared so little about education in the past that they have never spent enough money either on schools or on salaries of teachers to cater adequately for the nation as a whole. They have, of course, paid lip-service to the importance of education; but when it has come to producing enough money to develop a really effective and fair educational system, other demands on the Exchequer have always seemed more pressing. The reply to this is obvious enough—that radical improvements will only be effected in the state system when the wealthier and more influential minority of the population have to send their children to the same schools as everybody else.

And so the debate proceeds over the status of independent schools with arguments deriving from liberty, equality, and the quality of life. It represents *par excellence* an issue where these basic principles conflict. Obviously some compromise which does not involve abandoning altogether the right of parents to educate their children as they think fit will have to be found. The point at which such a compromise can be found must depend on the ingenuity of practical men. It cannot be determined by philosophical analysis alone, though such analysis can do much to get clearer on what fundamental points decisions have to be made.

NOTES

1. Spinoza, *Ethics*, Part IV, Prop. XXXV. Corollaries 1 and 2.
2. See Mill, J. S., *On Liberty*, Ch. II.
3. See Popper, K., *The Open Society and Its Enemies* (London, Routledge, 1945), Vol. I, pp. 225–6.
4. Locke, J., *The Second Treatise on Civil* Government (Ed. Gough, J., Oxford, Blackwell, 1948). Ch. VI, Sec. 57.
5. For problems about such 'other ways' see Peters, R. S., 'The Paradox of Moral Education' in Niblett, W. R. (Ed.), *Moral Education in a Changing Society* (London, Faber, 1963).

8. Equality of Opportunity

Charles Frankel

* * *

II

So we come to the notion of equality of opportunity. For it is to this notion that philosophers and ordinary men have commonly turned to bail them out of the difficulties in which the ideal of equality, taken by itself, appears to land them. The central place that equality of opportunity occupies in the analysis and defense of equality is well illustrated in the following passage from Brian Barry's book, *Political Argument:*

> Those who wish to disparage the distributive principle of equality often seek to do so by suggesting that its adherents are committed to holding either that men *are* 'equal' in their personal characteristics or that they *ought* to be 'equal'. Then, since 'equality of personal characteristics' does not seem to make much sense it is suggested that equalitarians presumably mean 'identical' when they say 'equal'. As this idea is absurd, too, distributive equality can be conveniently dismissed as an unintelligible concept. . . . What equality 'really means' it is claimed is that some reason or other must be adduced to justify treating people differently. The incoherence, however, lies not in the concept of equality, but in the hostile formulation itself. To say that people should be equal is to say that their opportunities for satisfying whatever wants they may happen to have should be equal. Whether or not one agrees with the claim in any particular case, it surely cannot be denied that it is a reasonably intelligible one, and one not involving any implausible prescriptions or descriptions involving uniformity or identity.[1]

To invoke the notion of equality of opportunity seems to be to take care of a number of problems. It takes account of the diversity of human wants and

Reprinted with permission of the publisher (THE UNIVERSITY OF CHICAGO PRESS) from *Ethics* 81 (April 1971): 199–211.

capacities. It answers the question as to just what it is, among all the possible goods of life, that should be equally distributed. And it explains both why inequality and difference may be accepted and what the limits to such acceptance should be in a just society. It thus appears to fill major gaps in the argument for equality.

Let us turn, therefore, and inspect this idea. Does it in fact perform these services? If we take Mr. Barry's formulation of the idea, it does so only on condition that we do not take it quite literally. We must read certain limitations into it. "To say that people should be equal," according to Barry, "is to say that their opportunities for satisfying whatever wants they may happen to have should be equal." He goes on to say that, even if we do not agree with this claim, it is at least "reasonably intelligible," and does not involve "any implausible prescriptions or descriptions involving uniformity or identity." But this is so, I believe, only if we silently decide that it does not mean certain things.

1. First of all, we would have to mean only *legitimate* wants. It would be quite implausible to argue, in the name of equality of opportunity, that a man who wants to torture all people over thirty ought to have an equal opportunity to satisfy this want with a man who wants, say, to support his aged parents. I recognize, to be sure, that moral standards are changing, so that, for all I know, this example may be infelicitous. Still, I take it that no one who says that men should have equal opportunities to satisfy whatever wants they may happen to have really means "whatever wants" without qualification. We cannot accept equality of opportunity unless we accept certain moral standards, above and beyond it, which limit its field of operation.

2. The restrictions which we must silently read into the conception of equality of opportunity go beyond the notion of morally legitimate wants. It is possible to have wants that are not in themselves morally illegitimate, but that are unrealistic. A man may want to be universally liked, he may want an economic system about which no one at all will complain; he may want a university which devotes its major energies to political activity, but which is at the same time a congenial place for skeptical dialogue and purely theoretical inquiries. Clearly, no human arrangement can give equality of opportunity to satisfy such wants. We implicitly exclude from the list of wants that people ought to have an opportunity to satisfy, I assume, those wants which no one can satisfy. Not that this prevents people from having such wants, from pushing for their satisfaction, or from demanding equality of opportunity to do so. But demands for the satisfaction of such unrealistic wants are nevertheless unacceptable. Ought implies can, and equality of opportunity does not provide an escape clause from this maxim.

3. These restrictions on equality of opportunity are fairly obvious. However, they lead to another which is less so and which suggests that the

concept of equality of opportunity is somewhat more puzzling, or at any rate more complicated, than Barry's words suggest.

There are many human wants which are not unrealistic in the absolute sense we have described: conditions can be defined, that is to say, under which they could be satisfied, and these conditions are not impossible to create. However, it would be extremely costly to do so. To take an example which, unfortunately, is not at all hypothetical, consider the supersonic transport plane. It is easily possible to create conditions in which people who want to fly from New York to California in two hours will have an opportunity to do so equal to the opportunity of those who are content to make the journey in five. But is it worth it? What is the cost to the nervous systems of people on the ground, and to other social needs which have been subordinated to this one? And obviously, many other examples can be given, a good portion of them no more imaginary.

It seems to me highly doubtful, therefore, that anyone who asks for equality of opportunity can consistently mean to say that he wants a society in which people's opportunities for satisfying whatever wants they may happen to have will be equal. Human wants conflict; they are multiple and insatiable; resources, if only the resources of human time and energy, are always scarce in relation to them. Some general system of social cost accounting, which assigns different values to the satisfaction of different wants, therefore has to be employed. This represents a substantial limitation on the ideal of equality of opportunity.

4. In addition to these qualifications that have to be introduced into the conception of equality of opportunity, there is still another issue. The idea as it is normally used is ambiguous. It points in two directions. Sometimes we invoke it to condemn a situation in which people are unable to satisfy their wants, but sometimes we invoke it to condemn a situation in which they are satisfying the wrong wants, or not sufficiently ambitious ones.

Thus, it is easier for the child of a working-class family to drop out of school than for a middle-class child; but the working-class child does not suffer from inequality of opportunity in the sense we have so far been discussing it, because dropping out of school is what he wants to do. In fact, it is the middle-class child who is more likely to be suffering from inequality so defined, since there are many more in this category who want to leave school but are unable to do so. Yet most egalitarians offer such facts as these as evidence of inequality of opportunity for workers and hold that in some way it should be rectified. Their complaint, therefore, is not about unequal opportunity to satisfy the wants that people happen to have; it is about unequal opportunity to develop the right wants. Eliza Doolittle had no desire to speak the king's English; she was perfectly content speaking the English she did until Professor Higgins got hold of her. But this is precisely what proves that she did not have equality of opportunity with Professor

Higgins. So there are not only moral standards and considerations of social cost that affect our notion of equality of opportunity. Cultural standards and notions of human potentialities may also be part of it.

Nor can it be argued, so far as I can see, that the notion of equality of opportunity cannot or should not be extended in this way. We do not condemn oppressive environments only because people feel oppressed inside them; we condemn them even when their victims do not feel oppressed. We do so because we think that it adds to the evil of an environment that it crushes people's power to imagine other possibilities and renders them wholly accepting of their condition. Equality of opportunity, in consequence, may quite properly lean on a conception of human or social excellence; but when it does, we cannot evaluate the demand for equality of opportunity without evaluating that conception. And it becomes different from the flatter conception of equal opportunity, conceived as opportunity to satisfy existing wants.

5. If ambiguities arise when we look at "wants," equivalent ambiguities emerge when we focus on the notion of "opportunity." Indeed, its ambiguity explains, I believe, some of our bitterest social controversies.

An example will help to bring out this ambiguity. Most people would agree, I think, that it would be odd for me to complain that I never had an equal opportunity with Mickey Mantle to play center field for the Yankees. Of course, I never did have an equal chance. But the competition for center fielder of the Yankees is perfectly fair; it is based on a test of capacities for the position and nothing else; anyone may enter the competition, and, in fact, great efforts are made to see that everyone qualified does enter; and, apart from unforeseeable accidents, the only thing that makes a difference is a man's ability. All that separates me from Mickey Mantle is my inability to compete with him, and since this is all that separates me, I may complain about my fate, but I cannot complain of inequality of opportunity. For when we speak of equality of opportunity to achieve something we set a man's abilities aside in estimating his chances and refer only to his chances to use them.

But much depends, therefore, on the notion of "ability." And here there is an equivocation. Suppose it were the case, to return to my disappointment, that I really could have competed successfully with Mickey Mantle, but that, being a city boy, I was discouraged from a very early age from doing so. My parents put other goals before me; there was not enough open space to play; my companions were an improperly motivated group who never offered me competition sufficient to challenge me, and who, in fact, often preferred to read books. Had these circumstances been different, I would have given Mickey Mantle a hard time. I had the native ability, and I was simply a victim of circumstance. If all this were the case, why could I not complain that I never had equality of opportunity with Mickey Mantle?

Why, indeed, do people so tamely accept the proposition that I did have an equal chance and that the best man simply won?

We begin to see what has happened here in this extension of this example, and in my movement from a mood of resignation to one of rebellion, by noticing that we use the term "ability" in at least two different contexts. There are contexts in which the primary desideratum is performance, here and now. In these contexts we use the term "ability" to refer to a man's general quality of performance. And in these terms, I had an equal opportunity to show my ability with Mickey Mantle. But there are other contexts in which the primary desideratum is developmental, educational, the evoking of potentialities. And in these contexts—for example, in schools—we commonly distinguish between an individual's "natural ability," as revealed by diagnostic tests or other means, and his actual performance. And when he performs at a level lower than his abilities we explain this in terms of his environment, or motivation, or physical health, or some other factor presumably extraneous to his ability. All these factors, of course, have something to do with his total performance; they make him, in fact, unable to do better than he is doing. Yet we do not say that he does not have the ability. For in these educational contexts what we do is to distinguish between two kinds of factors involved in performance, one which is modifiable and the other which is not, except within narrow limits. And we call the latter kind of factor "natural ability."

Thus, just as there arise different practical conceptions of equality of opportunity depending on whether we are talking about wants as they exist or wants as they should be, so there also arise different conceptions of equality of opportunity depending on whether we are stressing performance or the development of individual potentialities. If we stress the former, we arrive at what may be called the "meritocratic" conception of equality of opportunity. It holds that tests should be fair, that they should be open to everyone, that lack of money or other physical hindrances should not be a barrier to taking them, and that people should then be graded and rewarded in terms of their performance. So interpreted, equality of opportunity is entirely compatible with sharp hierarchical differences in society so long as their is also social mobility. It simply consists in the claim that social differentiation should be based on reasonable and objective principles, and that individuals should move up and down the hierarchy in accordance with their performance. It says nothing about the need to eliminate sharp distinctions, except insofar as this may be necessary to give everybody the same chance to compete.

Yet, clear as this "meritocratic" conception seems, it fades at the edges when it is pushed. If it seems unfair—an inequality of opportunity—that a man should not be able to take a test for a position he wants because he cannot afford to travel to the testing place—and most advocates of meritocracy would accept this as unfair—why is it not also unfair for a man to be deprived of the opportunity to prepare himself for such a test because he cannot get the necessary education? And if it is a mark of unequal opportunity to allow a man to be deprived of an education from which he would benefit when such an education is available to others, why is it not equally an example of unequal opportunity to leave him in an environment that deprives him even of the desire to seek such an education?

Thus, we move gradually to another conception of equality of opportunity—what I think might be called the "educational" conception. It looks upon the meritocratic approach as stilted, narrow, and coldly artificial. It condemns it for taking people simply as they are, for judging them in terms of their performance without asking what it is that makes one man perform better than another. Equality of opportunity means that men shall not be limited except by their abilities; the advocate of the "educational" conception of equality of opportunity holds that we cannot have real equality of opportunity unless we successfully modify those aspects of the individual's situation which prevent him from performing up to the level of his natural abilities.

III

Is there any way of adjudicating between these two conceptions? Not, so far as I can see, in a wholesale manner. There is no formula that allows us to settle the matter a priori. We have to proceed case by case. For, if I am right in saying that these two versions of equality of opportunity emerge out of different contexts in which different purposes are primary, then each represents a potentially legitimate claim in the making of public policy. There are circumstances in which what we want and cannot compromise with is performance; we cannot put up with inferior airline pilots or brain surgeons on the ground that they are learning on the job. There are other circumstances in which searching for and nourishing talent is the primary requirement; we would think it foolish to tell an eight-year-old chess player that he could not play again because he had lost a game to a chess master. And in between, there are all sorts of circumstances in which a concern for efficiency and a concern for education are both possible. In such circumstances we have to decide how much weight we shall give to each.

However, if there is no defensible general formula that allows us to come down neatly on one side or the other, there are certain guidelines, I believe, that help us to adjudicate specific cases. I shall suggest what I think the principal ones are, for they also help us to see, I think, what the logic of the case for equality of opportunity, in either its meritocratic or educational version, is.

1. One guideline is cost—not only economic cost narrowly considered, but economic cost in the broader terms of human time, energy, striving, and the probabilities of disappointment as against success. Let us go back to what we have noticed about the "educational" interpretation of equality of opportunity. It rests, we have seen, on a distinction between the individual's "natural" or "inherent" abilities, which are not subject to significant modification, and other factors which are held to be modifiable. But terms like "modifiable" and "unmodifiable" very often express only a difference in degree, and not a difference in kind. What is the cost, for example, of changing an individual's early environment as against, say, changing his genetic constitution? In the light of new developments in biology, it may some day be easier and less costly to change the genetic constitution of individuals. The language of "natural abilities" is likely in these circumstances to become even fuzzier than it now is. It is possible that we shall begin to talk of an individual's "natural abilities" in terms of those aspects of his personality which are determined by his early upbringing, rather than in terms of those that are genetically determined.

More to the immediate point, this example brings out the fact that while, in abstract principle, environmental factors are subject to modification, it is often extremely difficult to do so in practice. How do government, or the school, or organized psychological counseling, successfully intervene in methods of child-rearing affecting, for instance, the first three crucial months of the individual's life? How can the habits, fears, images of authority, and unconscious drives of the mother—or of the stand-in parents in the public institutions that might be created—be effectively changed? Despite extravagant claims, we do not really know very much about how to do this in a practical way. And what we do know indicates that it probably lies beyond our existing resources in people, funds, and general patience and goodwill. Certain limits, therefore, have to be placed on the applicability of the broad "educational" version of equality of opportunity. We can adopt it only to the extent that we can envisage circumstances that are, for practical purposes, modifiable, and that we think are worth modifying, given the cost.

There are, therefore, general differences in the way in which poor and rich societies normally construe equality of opportunity. In a poor society, at any rate one seriously committed to escaping from poverty, successful performance is urgently needed and false steps are costly. It is natural, therefore, that a meritocratic version of equality of opportunity usually takes hold in such societies, and that there should be a tendency in them to act on the principle that the race should go to the swift. This has happened in socialist societies as well as in nonsocialist ones. In richer societies, in contrast, the ampler, more educational view of equality of opportunity has a better chance to prevail. Even in a rich society, however, there are limits. If, for example, a society is able to obtain an adequate supply of first-rate mathematicians by relying on those who emerge, with no special educational effort, from the more favored classes of the population, how much should it expend to change the environment of poor people so that they make a proportionate contribution of mathematicians? The answer to this question is an open one. It involves the specification—a very difficult one—of a variety of costs on both sides of the issue: on one side, for example, the cost of maintaining an opportunity structure that limits horizons, or the cost of a socially insulated scientific elite; on the other side, the cost of undertaking experiments that may not work, or the limits on the need of the society for mathematicians, etc. These are very difficult issues to weigh, but they are examples of the kind of issue that has to be weighed when we try to determine whether the educational or meritocratic version of equality of opportunity should be stressed in a particular situation.

2. A second issue that emerges, as may already be evident, is an issue of morals, or at least of mores. The question of what is or is not modifiable, of what is or is not "natural ability," turns, in part, on what men choose to think should be modified, and what they place beyond the realm of organized social attack. As Plato pointed out, if we really want everyone to begin at the same place, so that only "natural abilities" control the outcome, there is one thing we must absolutely do—abolish the family and bring people up in public institutions. The family makes a more immediate difference than anything else in determining the individual's life chances. Unless we are ready to deny the institution of the family the special protections we now assign to it, only some of the inequalities associated with family origins (and, exceptional cases apart, probably not the most important) can be changed. The justification of any particular demand for equality of opportunity in the broad nonmeritocratic sense depends on whether the demand touches on fundamental matters of this sort and on the degree to which we are willing and able to do something about them.

3. A third issue has to do with the standards by which we measure either performance or ability. Let us take as an example the volatile issue of equality of educational opportunity. It has been said with increasing frequency in Europe recently in relation to the de facto segregation of social classes in the schools, and with equal or greater frequency in the United States in relation to our racial problems, that equality of educational opportunity is in effect denied because, in measuring "ability" or "performance," we do so by standards that are culturally biased: they are bourgeois, or white middle class, or somthing of the sort. Accordingly, those coming from other milieux cannot usually compete successfully; and, in any case, it is an imposition to ask them to compete, since the standards in question are

legitimate only in terms of social purposes that are not theirs. Equality of opportunity therefore entails, it is argued, changing the nature of the standards applied to those who are disadvantaged.

Here again the issue is one, I believe, of striking a balance. Where the needs of people in different social groups are different, and where their aptitudes lie in different areas, there is a legitimate claim that standards should be modified to take these into account. However, the degree of legitimacy of this claim is limited by the answers we give to other questions: To what extent are we willing to create an educational system that will fix most people in separate social or ethnic groups? And to what extent are the needs and aptitudes we are asked to take into account needs and aptitudes that, in the long-run, are socially and historically viable? For the argument that standards should be modified for the sole and sufficient reason that they do not fit the distinctive situation of a given group rests on at least two fallacies.

It assumes, to begin with, that all the standards employed are culturally biased. But this is not so unless we are prepared to say that there is such a thing as bourgeois mathematical logic or white middle-class astronomy. At least some intellectual disciplines are cross-cultural, and standards of performance or ability developed in relation to them are also cross-cultural in validity, even if not in their origins. Moreover, the claim that because workers and bourgeois, or blacks and whites, differ in cultural tastes and aptitudes in certain respects, they differ in all respects, is a non sequitur, and plainly an exaggeration.

Second, even if we agree that many tests of school performance or ability are culturally biased, which I think we must admit, this does not prove that individuals coming from different cultural backgrounds should not be measured by them. It is possible that they should be measured by them in their own interest. Ability in arithmetic, for example, is a bourgeois value. But if ability in arithmetic is useful to any citizen of an industrial society, then it is a reasonable standard to expect any citizen to meet. This does not preclude also varying educational requirements where the long-range needs of different individuals or groups are different. Nor does it preclude the criticism and correction of educational standards when these are culturally biased to no good point. But equality of opportunity remains a demand we can evaluate only if we ask: Equality of opportunity for what?

IV

It is tempting to stop here, for this essay is already long. However, if I were to do so, I would stop before a final and crucial feature of the ideal of equality of opportunity had been discussed. It is a feature that the ideal of equality of opportunity shares with all other social ideals. I have said that the merits of applying it, and the kind of interpretation we give it in particular cases, depend on the nature of the particular cases and on the answers we give to a broad array of highly complicated factual and moral questions. In relation to making specific decisions about public policy, that is perhaps all that needs to be said. However, decision making goes on in a broader environment, which puts pressure on decision makers and affects the general trend of their decisions. And in this broader environment, generalized social ideals, not carefully modified and qualified to take account of different individual cases, play a crucial role. So we cannot ask about a social ideal simply what it means in particular cases; we have to ask what its general tendency to influence policy is.

More specifically, an ideal like equality of opportunity serves two broad and general purposes above and beyond functioning as a claim that has to be adjudicated in specific contexts. It serves, first, as a rough-and-ready rule of thumb which gives an initial bias to the answers decision makers give and which they use because they require such a bias. One need only recall the sort of questions which, if my analysis has been right, must be asked when specific decisions about equality of opportunity are made. They are questions inviting attention to such a broad range of facts and values and raising so many issues to which the answers must in part be speculative that it is difficult to see how any determinate answer can be given to them, unless, on principle, we are inclined to lean in one direction or another and to take a chance on one hypothesis rather than another. And this in fact we do, and have no alternative but to do. For we can suspend belief when we do not know enough, but we cannot suspend decision. And so, consciously or unconsciously, we have to make certain general decisions of principle—not decisions about unbreakable or absolute principles, but decisions about guiding principles. The principle (or principles) of equality of opportunity is an answer to such a requirement. It loads the dice for us, because we need the dice loaded.

The second function of such an ideal is a related one. Public decisions are made in situations marked by massive, conflicting pressures. Established positions resist decisions that strike out along new paths; established habits of thought pose the issues in one way and ignore facts and possibilities that can be seen from other perspectives. On the other side, new congeries of power and interest push out in other directions; and new ideas function to weaken the hold of inherited habits. Decisions are made in these circumstances; and much depends, therefore, on the general drift of sentiment and aspiration and on the ways in which it comes to seem legitimate initially to pose an issue. It is this that can be affected by broad social ideals and by the decisions we make, as philosophers, educators, lawyers, or ordinary citizens, to support one general ideal as against another.

For social principles have a tendency to spread out, to spill over into areas different from those in which they were generated, to raise analogies where they were not suspected before, and, in general, to make trouble where it was not contemplated that they would. We may say that this comes from interpreting these ideals loosely and forgetting the circumstances with respect to which they were developed. Perhaps so; but the fact remains that this is what happens, and so we have to take it into account when we formulate or promulgate an ideal. The ideal that wins out tells us what general sort of question to ask of the status quo, what general sort of moral pressure to put it under, in what direction, as a broad matter, we should try to move it. Accordingly, beyond asking what a particular ideal means in particular contexts of decision making, we have to ask whether we are willing to live with it as a not quite tamed, freely roaming creature whose existence affects the general atmosphere.

If we raise these very general considerations, my own sympathies go to the broader, "educational" view of equality of opportunity. Its function is to call attention, at least indirectly, to the fact that situations with educational (or miseducational) components occur much more frequently than is usually recognized, and that more weight should be given to the educational aspects of such situations than has been given in the past. Part of the logic of such an assertion is, presumably, that this will also, in the long run, increase the efficiency of performance in the society at large. But another of the reasons for betting on it is that it implicitly proposes that we consider other values besides industrial productivity, narrowly conceived, in measuring the worth of a society. It asks us to consider the impact of the society on the formation and development of personality. It proposes a humanistic and not a technological notion of efficiency.

This does not mean that we need formulate this ideal, even for general purposes, in unguarded terms. Social ideals that unloose overreaching expectations have cruel consequences. It is clearly unacceptable to adopt the ideal of "educationl" equality of opportunity in the extreme form in which it is sometimes stated, so that all differences in the average achievement of different social classes are put down to remediable conditions which the political process, the courts, other organized social agencies, or perhaps a revolutionary movement, have an obligation to remove. The ideal of "educational" equality of opportunity should be taken to state a direction of effort, not a goal to be fully achieved. In education itself, for example, it cannot be taken to call for equality of achievement by all individuals, but only for comparable levels of average achievement in different social classes.[2] And even this can only be rough comparability, a matter of more or less, for it is not possible for schools to counteract entirely the differential influence of specific environments, nor does any society know how to do this except within certain limits.[3]

As a practical matter, therefore, "equality of opportunity" calls not for uniformity, either of environment or achievement. It calls for the diversification of opportunities, the individualization of attention in schools and work places, the creation of conditions making it easier for people to shift directions and try themselves out in new jobs or new milieux, and a general atmosphere of tolerance for a plurality of value-schemes insofar as this is feasible. Such a practical policy goes beyond the narrow meritocratic conception. It would require, and it would presumably lead to, a greater equalizing of social conditions. But it would not promise a state of affairs in which it was just as easy for those less favored by circumstance as for those more favored to satisfy whatever wants they may happen to have.

A man does not have to be poor to be disadvantaged; he merely needs to be poorer than somebody else. And while we can eliminate differences in pecuniary income if we decide to, we cannot eliminate other important differences in circumstances unless we wish to adopt the principle that parents should not feel any special devotion to their own children and should be prevented from passing on to them what advantages in motivation, knowledge, or personal associations they may happen to possess. I am inclined to think this is not a practical principle. I am even more persuaded that it is not a desirable one. So equality of opportunity, as a matter of policy, should aim at striking a mean between the "meritocratic" and "educational" versions of the ideal. But, to maintain the Aristotelian analogy, it should lean a bit in one direction; that direction, I think, should be toward the broader "educational" version.

But why care about "equality of opportunity" at all? Why care about "equality"? At the level of broad and general choice between ideals, why make this choice? I come here to my concluding remarks, and they can only suggest another essay; they cannot be that essay. So I will only say, without the argument that is required, that the case for equality does not seem to me to be a demonstrative case. It does not follow deductively from any first principles. It comes from the connection of equality with a whole cluster of other values. It comes from its practical implications, if we believe in it and act on it, for our other attitudes. The case for equality is not equality in itself. It is the value of liberty, diversity, and, most of all, fraternity. Within broad limits (of which we should try to be aware) equality promotes these values.

Antiegalitarians from Plato to Mencken have alleged that the demand for "equality" is often only the disguised expression of envy, the inferior man's way of taking revenge on his betters. They have said that the pursuit of distributive equality can lead to suspicion toward distinction and hostility to firm standards. Tocqueville further observed that it can lead to the loneliness, anxiety, and pressures to conformity that mark societies in which class lines are vague and individuals cannot be sure where they belong or who are their kind of people.

I think there is some truth in these arguments. Yet those who have made them have almost invariably been unfair. They have taken egalitarianism as a free-roaming social ideal and compared it in its worst manifestations with aristocratic notions at their best. They have overlooked the fact that the latter may also run out of control and usually do, and that, under these circumstances, they stand, not for recognition of excellence, but for privilege, oligarchy, and fear of new forms of human achievement. We must compare egalitarians at their worst with the aristocrats whom Milton described as "drunk on wine and insolence."

Moreover, the opponents of egalitarianism, with the prominent exception of Tocqueville, have missed its special grace and charm. It can, at its best, make all men the objects of a common friendly regard. And even at its middling best, when it is pushing and competitive, it produces the sense that doors are open and that no one need be shut out. It thus gives practical substance to guarantees of liberty, and at least one kind of reality to the hope for fraternity. And by encouraging people with talent to think they have a chance to use it, it probably contributes to the general development of talent in society. Not least, though it encourages hostility toward those who stand out from the crowd, it also invites a mixing of human types and a steady challenge to conventional standards of ability and achievement that bring excitement and variety to human experience.

Indeed, although the egalitarian is often considered excessively worldly, and too much focused on material concerns, there is an other-worldly aspect to egalitarianism. It looks ironically on worldly distinctions; it pronounces all titles, ranks, and stigmata of achievement to be things of limited and equivocal significance; it is the enemy of pomposity. Of all social outlooks, it is, therefore, the most congenial, probably, to the flowering of compassion.

NOTES

1. Brian Barry, *Political Argument* (New York, 1965), p. 120.
2. See James S. Coleman, "The Concept of Equality of Educational Opportunity," in *Equal Opportunity*, ed. *Harvard Educational Review* staff (Cambridge, Mass.: Harvard University Press, 1969).
3. For example, if new methods of instruction are discovered which make it easier to raise students' reading capacities, the more favored classes

will benefit as much as the poorer classes, and perhaps more, since they are likely to be able to exploit these methods sooner and to have less resistance to them. Thus, the gap between the richer and poorer classes in reading ability may remain, even though the average for the whole society has been raised.

9. Moral Education and the Democratic Ideal

Israel Scheffler

INTRODUCTION

What should be the purpose and content of an educational system in a democratic society, in so far as it relates to moral concerns? This is a very large question, with many and diverse ramifications. Only its broadest aspects can here be treated, but a broad treatment, though it must ignore detail, may still be useful in orienting our thought and highlighting fundamental distinctions and priorities.

EDUCATION IN A DEMOCRACY

Commitment to the ideal of democracy as an organizing principle of society has radical and far-reaching consequences, not only for basic political and legal institutions, but also for the educational conceptions that guide the

Originally prepared at the invitation of Representative John Brademas, Chairman, Select Subcommittee on Education, Committee on Education and Labor of the House of Representatives. It was offered as a background paper for the Subcommittee's hearings on the proposed National Institute of Education. A version also appears in *Educational Research: Prospects and Priorities*, Appendix 1 to Hearings on H.R. 3606, Committee Print, 92nd Congress, 2nd Session, Washington: US Government Printing Office, 1972. Written in response to a growing interest in moral education as a special area or aspect of schooling, the paper stresses the connections between moral, scientific, and democratic education and the centrality, in all three, of the habits of critical thought. (Extracts from R. B. Perry, *Realms of Value*, 1954, by permission of Harvard University Press.)

Reprinted with permission of the author from *Reason and Teaching* (Indianapolis, Ind.: The Bobbs-Merrill Company, Inc., 1973, copyright © 1973 by Israel Scheffler), pp. 136–145.

development of our children. All institutions, indeed, operate through the instrumentality of persons; social arrangements are 'mechanisms' only in a misleading metaphorical sense. In so far as education is considered broadly, as embracing all those processes through which a society's persons are developed, it is thus of fundamental import for all the institutions of society, without exception. A society committed to the democratic ideal is one that makes peculiarly difficult and challenging demands of its members; it accordingly also makes stringent demands of those processes through which its members are educated.

What is the democratic ideal, then, as a principle of social organization? It aims so to structure the arrangements of society as to rest them ultimately upon the freely given consent of its members. Such an aim requires the institutionalization of reasoned procedures for the critical and public review of policy; it demands that judgments of policy be viewed not as the fixed privilege of any class or élite but as the common task of all, and it requires the supplanting of arbitrary and violent alteration of policy with institutionally channeled change ordered by reasoned persuasion and informed consent.

The democratic ideal is that of an open and dynamic society: open, in that there is no antecedent social blueprint which is itself to be taken as a dogma immune to critical evaluation in the public forum; dynamic, in that its fundamental institutions are not designed to arrest change but to order and channel it by exposing it to public scrutiny and resting it ultimately upon the choices of its members. The democratic ideal is antithetical to the notion of a fixed class of rulers, with privileges resting upon social myths which it is forbidden to question. It envisions rather a society that sustains itself not by the indoctrination of myth, but by the reasoned choices of its citizens, who continue to favor it in the light of the critical scrutiny both of it and its alternatives. Choice of the democratic ideal rests upon the hope that this ideal will be sustained and strengthened by critical and responsible inquiry into the truth about social matters. The democratic faith consists not in a dogma, but in a reasonable trust that unfettered inquiry and free choice will themselves be chosen, and chosen again, by free and informed men.

The demands made upon education in accord with the democratic ideal are stringent indeed; yet these demands are not ancillary but essential to it. As Ralph Barton Perry has said:[1]

> Education is not merely a boon conferred by democracy, but a condition of its survival and of its becoming that which it undertakes to be. Democracy is that form of social organization which most depends on personal character and moral autonomy. The members of a democratic society cannot be the wards of their betters; for there is no class of betters . . . Democracy demands of every man what in other forms of social organization is demanded only of a segment of

society . . . Democratic education is therefore a peculiarly ambitious education. It does not educate men for prescribed places in life, shaping them to fit the requirements of a preexisting and rigid division of labor. Its idea is that the social system itself, which determines what places there are to fill, shall be created by the men who fill them. It is true that in order to live and to live effectively men must be adapted to their social environment, but only in order that they may in the long run adapt that environment to themselves. Men are not building materials to be fitted to a preestablished order, but are themselves the architects of order. They are not forced into Procrustean beds, but themselves design the beds in which they lie. Such figures of speech symbolize the underlying moral goal of democracy as a society in which the social whole justifies itself to its personal members.

To see how radical such a vision is in human history, we have only to reflect how differently education has been conceived. In traditional authoritarian societies education has typically been thought to be a process of perpetuating the received lore, considered to embody the central doctrines upon which human arrangements were based. These doctrines were to be inculcated through education; they were not to be questioned. Since, however, a division between the rulers and the ruled was fundamental in such societies, the education of governing élites was sharply differentiated from the training and opinion-formation reserved for the masses. Plato's *Republic*, the chief work of educational philosophy in our ancient literature, outlines an education for the rulers in a hierarchical utopia in which the rest of the members are to be deliberately nourished on myths. And an authoritative contemporary Soviet textbook on *Pedagogy* declares that, 'Education in the USSR is a weapon for strengthening the Soviet state and the building of a classless society . . . the work of the school is carried on by specially trained people who are guided by the state.'[2] The school was indeed defined by the party program of March 1919 as 'an instrument of the class struggle. It was not only to teach the general principles of communism but "to transmit the spiritual, organizational, and educative influence of the proletariat to the half- and nonproletarian strata of the working masses."'[3] In nondemocratic societies, education is two faced: it is a weapon or an instrument for shaping the minds of the ruled in accord with the favored and dogmatic myth of the rulers; it is, however, for the latter, an induction into the prerogatives and arts of rule, including the arts of manipulating the opinions of the masses.

To choose the democratic ideal for society is wholly to reject the conception of education as an *instrument* of rule; it is to surrender the idea of shaping or molding the mind of the pupil. The function of education in a democracy is rather to liberate the mind, strengthen its critical powers, inform it with knowledge and the capacity for independent inquiry, engage its human sympathies, and illuminate its moral and practical choices. This function is, further, not to be limited to any given subclass of members, but to be extended, in so far as possible, to all citizens, since all are called upon to take part in processes of debate, criticism, choice, and co-operative effort upon which the common social structure depends. 'A democracy which educates for democracy is bound to regard all of its members as heirs who must so far as possible be qualified to enter into their birthright.'[4]

IMPLICATIONS FOR SCHOOLING

Education, in its broad sense, is more comprehensive than schooling, since it encompasses all those processes through which a society's members are developed. Indeed, all institutions influence the development of persons working within, or affected by, them. Institutions are complex structures of actions and expectations, and to live within their scope is to order one's own actions and expectations in a manner that is modified, directly or subtly, by that fact. Democratic institutions, in particular, requiring as they do the engagement and active concern of all citizens, constitute profoundly educative resources. It is important to note this fact in connection with our theme, for it suggests that formal agencies of schooling do not, and cannot, carry the whole burden of education in a democratic society, in particular moral and character education. All institutions have an educational side, no matter what their primary functions may be. The question of moral education in a democracy must accordingly be raised not only within the scope of the classroom but also within the several realms of institutional conduct. Are political policies and arrangements genuinely open to rational scrutiny and public control? Do the courts and agencies of government operate fairly? What standards of service and integrity are prevalent in public offices? Does the level of political debate meet appropriate requirements of candor and logical argument? Do journalism and the mass media expose facts and alternatives, or appeal to fads and emotionalism? These and many other allied questions pertain to the status of moral education within a democratic society. To take them seriously is to recognize that moral education presents a challenge not only to the schools, but also to every other institution of society.

Yet the issue must certainly be raised specifically in connection with schools and schooling. What is the province of morality in the school, particularly the democratic school? Can morality conceivably be construed as a *subject*, consisting in a set of maxims of conduct, or an account of current mores, or a list of rules derived from some authoritative source? Is the function of moral education rather to ensure conformity to a certain code of behavior regulating the school? Is it, perhaps, to involve pupils in the activities of student organizations or in discussion of 'the problems of democracy'? Or, since morality pertains to the whole of what transpires in school, is the very notion of specific moral schooling altogether misguided?

These questions are very difficult, not only as matters of implementation, but also in theory. For it can hardly be said that there is firm agreement among moralists and educators as to the content and scope of morality. Yet the tradition of moral philosophy reveals a sense of morality as a comprehensive institution over and beyond particular moral codes, which seems to me especially consonant with the democratic ideal, and can, at least in outline, be profitably explored in the context of schooling. What is this sense?

It may perhaps be initially perceived by attention to the language of moral judgment. To say that an action is 'right,' or that some course 'ought' to be followed, is not simply to express one's taste or preference; it is also to make a claim. It is to convey that the judgment is backed by reasons, and it is further to invite discussions of such reasons. It is, finally, to suggest that these reasons will be found compelling when looked at impartially and objectively, that is to say, taking all relevant facts and interests into account and judging the matter as fairly as possible. To make a moral claim is, typically, to rule out the simple expression of feelings, the mere giving of commands, or the mere citation of authorities. It is to commit oneself, at least in principle, to the 'moral point of view,' that is, to the claim that one's recommended course has a point which can be clearly seen if one takes the trouble to survey the situation comprehensively, with impartial and sympathetic consideration of the interests at stake, and with respect for the persons involved in the issue. The details vary in different philosophical accounts, but the broad outlines are generally acknowledged by contemporary moral theorists.[5]

If morality can be thus described, as an institution, then it is clear that we err if we confuse our allegiance to any particular code with our commitment to this institution; we err in mistaking our prevalent code for the *moral point of view* itself. Of course, we typically hold our code to be justifiable from the moral point of view. However, if we are truly committed to the latter, we must allow the possibility that further consideration or new information or emergent human conditions may require revision in our code. The situation is perfectly analogous to the case of science education; we err if we confuse our allegiance to the current corpus of scientific doctrines with our commitment to scientific method. Of course we hold our current science to be justifiable by scientific method, but that very method itself commits us to hold contemporary doctrines fallible and revisable in the light of new arguments or new evidence that the future may bring to light. For scientific doctrines are not held simply as a matter of arbitrary preference; they are held for reasons. To affirm them is to invite all who are competent to survey these reasons and to judge the issues comprehensively and fairly on their merits.

Neither in the case of morality nor in that of science is it possible to convey the underlying *point of view* in the abstract. It would make no sense to say, 'Since our presently held science is likely to be revised for cause in the future, let us just teach scientific method and give up the teaching of content.' The content is important in and of itself, and as a basis for further development in the future. Moreover, one who knew nothing about specific materials of science in the concrete could have no conception of the import of an abstract and second-order scientific method. Nevertheless, it certainly does not follow that the method is of no consequence. On the contrary, to teach current science without any sense of the reasons that underlie it, and of the logical criteria by which it may itself be altered in the future, is to prevent its further intelligent development. Analogously, it makes no sense to say that we ought to teach the moral point of view in the abstract since our given practices are likely to call for change in the future. Given practices are indispensable, not only in organizing present energies, but in making future refinements and revisions possible. Moreover, one who had no concrete awareness of a given tradition of practice, who had no conception of what rule-governed conduct is, could hardly be expected to comprehend what the moral point of view might be, as a second order vantage point on practice. Nevertheless, it does not follow that the latter vantage point is insignificant. Indeed, it is fundamental in so far as we hold our given practices to be reasonable, that is, justifiable in principle upon fair and comprehensive survey of the facts and interests involved.

There is, then, a strong analogy between the moral and the scientific points of view, and it is no accident that we speak of reasons in both cases. We can be reasonable in matters of practice as well as in matters of theory. We can make a fair assessment of the evidence bearing on a hypothesis of fact, as we can make a fair disposition of interests in conflict. In either case, we are called upon to overcome our initial tendencies to self-assertiveness and partiality by a more fundamental allegiance to standards of reasonable judgment comprehensible to all who are competent to investigate the issues. In forming such an allegiance, we commit ourselves to the theoretical possibility that we may need to revise our current beliefs and practices as a consequence of 'listening to reason.' We reject arbitrariness in principle, and accept the responsibility of critical justification of our current doctrines and rules of conduct.

It is evident, moreover, that there is a close connection between the general concept of *reasonableness*, underlying the moral and the scientific points of view, and the democratic ideal. For the latter demands the institutionalization of 'appeals to reason' in the sphere of social conduct. In requiring that social policy be subject to open and public review, and institutionally revisable in the light of such review, the democratic ideal rejects the rule of dogma and of arbitrary authority as the ultimate arbiter of social conduct. In fundamental allegiance to channels of open debate, public review, rational persuasion and orderly change, a democratic society in effect holds its own current practices open to revision in the future. For it considers these practices to be not self-evident, or guaranteed by some fixed and higher authority, or decidable exclusively by some privileged élite, but subject to rational criticism, that is, purporting to sustain themselves in the process of free exchange of reasons in an attempt to reach a fair and comprehensive judgment.

Here, it seems to me, is the central connection between moral, scientific, and democratic education, and it is this central connection that provides, in my opinion, the basic clue for school practice. For what it suggests is that the fundamental trait to be encouraged is that of reasonableness. To cultivate this trait is to liberate the mind from dogmatic adherence to prevalent ideological fashions, as well as from the dictates of authority. For the rational mind is encouraged to go behind such fashions and dictates and to ask for their justifications, whether the issue be factual or practical. In training our students to reason we train them to be critical. We encourage them to ask questions, to look for evidence, to seek and scrutinize alternatives, to be critical of their own ideas as well as those of others. This educational course precludes taking schooling as an instrument for shaping their minds to a preconceived idea. For if they seek reasons, it is their evaluation of such reasons that will determine what ideas they eventually accept.

Such a direction in schooling is fraught with risk, for it means entrusting our current conceptions to the judgment of our pupils. In exposing these conceptions to their rational evaluation we are inviting them to see for themselves whether our conceptions are adequate, proper, fair. Such a risk is central to scientific education, where we deliberately subject our current theories to the test of continuous evaluation by future generations of our student-scientists. It is central also to our moral code, *in so far as* we ourselves take the moral point of view toward this code. And, finally, it is central to the democratic commitment which holds social policies to be continually open to free and public review. In sum, rationality liberates, but there is no liberty without risk.

Let no one, however, suppose that the liberating of minds is equivalent to freeing them from discipline. *Laissez-faire* is not the opposite of dogma. To be reasonable is a difficult achievement. The habit of reasonableness is not an airy abstract entity that can be skimmed off the concrete body of thought and practice. Consider again the case of science: scientific method can be learned only in and through its corpus of current materials. Reasonableness in science is an aspect or dimension of scientific tradition, and the body of the tradition is indispensable as a base for grasping this dimension. Science needs to be taught in such a way as to bring out this dimension as a consequence, but the consequence cannot be taken neat. Analogously for the art of moral choice: the moral point of view is attained, if at all, by acquiring a tradition of practice, embodied in rules and habits of conduct. Without a preliminary immersion in such a tradition—an appreciation of the import of its rules, obligations, rights, and demands—the concept of choice of actions and rules for oneself can hardly be achieved. Yet the prevalent tradition of practice can itself be taught in such a way as to encourage the ultimate attainment of a superordinate and comprehensive moral point of view.

The challenge of moral education is the challenge to develop critical thought in the sphere of practice and it is continuous with the challenge to develop critical thought in all aspects and phases of schooling. Moral schooling is not, therefore, a thing apart, something to be embodied in a list of maxims, something to be reckoned as simply another subject, or another activity, curricular or extracurricular. It does, indeed, have to pervade the *whole* of the school experience.

Nor is it thereby implied that moral education ought to concern itself solely with the general structure of this experience, or with the effectiveness of the total 'learning environment' in forming the child's habits. The critical questions concern the *quality* of the environment: what is the *nature* of the particular school experience, comprising content as well as structure? Does it liberate the child in the long run, as he grows to adulthood? Does it encourage respect for persons, and for the arguments and reasons offered in personal exchanges? Does it open itself to questioning and discussion? Does it provide the child with fundamental schooling in the traditions of reason, and the arts that are embodied therein? Does it, for example, encourage the development of linguistic and mathematical abilities, the capacity to read a page and follow an argument? Does it provide an exposure to the range of historical experience and the realms of personal and social life embodied in literature, the law, and the social sciences? Does it also provide an exposure to particular domains of scientific work in which the canons of logical reasoning and evidential deliberation may begin to be appreciated? Does it afford opportunity for individual initiative in reflective inquiry and practical projects? Does it provide a stable personal milieu in which the dignity of others and the variation of opinion may be appreciated, but in which a common and overriding love for truth and fairness may begin to be seen as binding oneself and one's fellows in a universal human community?

If the answer is negative, it matters not how effective the environment is in shaping concrete results in conduct. For the point of moral education in a democracy is antithetical to mere shaping. It is rather to liberate.

NOTES

1. Ralph Barton Perry, *Realms of Value*, Cambridge: Harvard University Press, 1954, pp. 431–2. Excerpt reprinted in I. Scheffler, ed., *Philosophy and Education*, 2nd ed., Boston: Allyn & Bacon. 1966, pp. 32 ff.
2. Cited in Introduction to George S. Counts and Nucia P. Lodge, eds and translators, *I Want To Be Like Stalin: From the Russian Text on Pedagogy* (by B. P. Yesipov and N. K. Goncharov), New York: John Day, 1947, pp. 14, 18. (The materials cited are from the 3rd ed. of the *Pedagogy*, published in 1946.)
3. Frederic Lilge, 'Lenin and the Politics of Education,' *Slavic Review* (June 1968), Vol. XXVII. No. 2, p. 255.
4. Ralph Barton Perry, *op. cit.*, p. 432.
5. See, for example, Kurt Baier, *The Moral Point of View*, Ithaca: Cornell University Press, 1958; William K. Frankena, *Ethics*, Englewood Cliffs, N.J.: Prentice Hall, 1963, and R. S. Peters, *Ethics and Education*, Glenview, Ill.: Scott Foresman, 1967. Additional articles of interest may be found in Sect. v, 'Moral Education' and Sect. vi, 'Education, Religion, and Politics,' in I. Scheffler, ed., *Philosophy and Education*.

Part Four: Value Judgments, Teaching, and the Curriculum

Introduction

One of the objectives of this anthology is to indicate the extent to which value judgments permeate the entire educational process. In that connection, the inextricable ties between values and schooling have seldom been demonstrated more cogently, yet concisely, than by John L. Childs in the first chapter of Part Four. In this brief excerpt from his book, *Education and Morals*, Childs points out that value judgments are an integral part of decisions affecting the curriculum, extracurricular activities, school regulations, grading, school-wide assemblies, school-community relationships, etc. Professor Childs uses the term "moral values" in reference to all of these contexts, while some writers would want to distinguish between moral and "nonmoral" values. But in either case, the point, of course, is that those who argue that schools should be value-free have simply failed to grasp what education is all about.[1]

In Chapter II, E. R. Emmet uses as example familiar to teachers everywhere—that of the schoolmaster faced with the task of grading a set of essays—to explore the nature of value judgments, with particular emphasis on the objectivity-subjectivity issue. Are value judgments similar to factual statements or do they more closely resemble expressions of taste? When we judge Helen's essay to be superior to Robert's, is the claim similar to the observation that one building is taller than another? Or is it like saying, "I prefer apples to oranges?" Is there a third alternative? Do value judgments fall somewhere in between statements of fact and declarations of personal preference? To put the question another way, are value judgments purely subjective, or is there some sense in which they might be said to include at least an element of objectivity? And if the latter, what criteria are appropriate in attempting to justify a value judgment? In the context of this essay, the justification issue, as Emmet notes, raises related questions of meaning, purpose, and expert opinion.

Of particular interest to educators is Emmet's discussion of expert opinion in relation to literature. Here again, the key question is whether or not the literary critic (or the professor of English) is expressing more than personal opinion when assigning literary merit to a given piece of writing. Of course expert opinion may also be attributed to the schoolmaster struggling with his essays. And if the judgments of both the English professor and the schoolmaster are entirely subjective, what are we to conclude with respect to our grading practices and selection of subject matter? To what extent, if any, can they be justified?

The problem of justification is addressed by Israel Scheffler in Chapter 12. Professor Scheffler considers two levels of justification: relative and general. Relative justification involves appeals to rules and may be compared to game activities, in which moves are limited by the formal rules of the game. In most real-life situations, however, justification is more complicated, since the appropriate norms are less evident than any given set of game rules. Moreover, relative justification will not suffice when we feel compelled to ask, (often with reference to many of our most important decisions) not simply whether an action is justifiable according to certain rules or practices, but whether it is justifiable, period. In such cases, we require general, rather than relative justification.

Consistency with a guiding set of rules is often a factor even in decisions involving general justification, but on occasion the rules themselves may be called into question. Attention should be called here to Professor Scheffler's comparison between the justification of "controllable acts" or "moves" and the acceptance of beliefs. His reference to formal logic as a codification of judgments regarding valid arguments is particularly helpful in illuminating these complex matters.

Scheffler's analysis of the two levels of justification is followed by a parallel consideration of educational decision making, with special emphasis on the justification of curricular decisions. At this point in the article, he lists several rules pertaining to such decisions and makes a number of specific recommendations regarding subject matter.

NOTES

1. As R. S. Peters has observed, "It would be as much of a logical contradiction to say that a person had been educated and yet the change was in no way desirable as it would be to say that he had been reformed and yet had made no change for the better." See R. S. Peters, "Education as Initiation," in *Philosophical Analysis and Education*, ed. Reginald D. Archambault (New York: The Humanities Press, 1965), pp. 90–91.

10. Characteristics of Deliberate Education

John L. Childs

* * *

THE MORAL NATURE OF DELIBERATE EDUCATION

As we have emphasized in all the foregoing, deliberate education is never morally neutral. A definite expression of preference for certain human ends, or values, is inherent in all efforts to guide the experience of the young. No human group would ever bother to found and maintain a system of schools were it not concerned to make of its children something other than they would become if left to themselves and their surroundings. Moreover, in order to develop the preferred and chosen patterns of behavior, it is necessary to hinder other and incompatible kinds of growth. A school is ineffective as an educational agency whenever the emphases in certain aspects or departments of its work are denied or negated in other parts of its program. In education, as in other realms of human activity, the actual practices of a school are more potent than its verbal professions. Maximum results are achieved when both the declared aims and the actual deeds of a school are unified, and its children are reared in an environment that supports in its daily practices that which it affirms in its theory.

As we have already stated, the term *moral*, as used in this discussion, does not pertain to a restricted phase of the work of the school. The moral interest pervades the entire educational program. It is involved whenever a significant choice has to be made between a better and a worse in the nurture of the young. The moral factor appears whenever the school, or the individual

Reprinted from *Education and Morals* (New York: John Wiley & Sons, Inc., 1967), pp. 17–20.

teacher or supervisor, is *for* certain things and *against* other things. The moral element is preëminently involved in all of those selections and rejections that are inescapable in the construction of the purposes and the curriculum of the school. It appears, for example, in the affairs of the playground—in the kind of sports that are favored and opposed, and in the code of sportsmanship by which the young are taught to govern their behavior in the actual play of the various games. It appears in the social life of the school—in all of the behaviors that are approved or disapproved as the young are taught the manners—the conventional or minor morals—of their society. It appears in the school's definition of the delinquent and in its mode of dealing with him. It appears in the way children are taught to treat those of different racial, religious, occupational, economic or national backgrounds. It appears in the department of science: in the methods the young are expected to adopt in conducting their experiments, in their reports of what actually happened during the course of their experiments, as well as in the regard of the teachers of science for accuracy, for precision, and for conclusions that are based on objective data rather than on wishful thinking. It appears in the department of social studies: in the problems that are chosen to be discussed, in the manner in which they are discussed, in the historical documents and events that are emphasized, as well as in the leaders that are chosen to illustrate the important and the worthy and the unimportant and the unworthy in the affairs of man. It appears in the department of literature: in the novels, the poems, the dramas that are chosen for study, in what is considered good and what is considered bad in the various forms and styles of human conduct and expression. It appears in the organization and the government of the school: in the part that superintendent, supervisors, teachers, pupils are expected to play in the making and the maintenance of the regulations of the school. It appears in the methods of grading, promoting, and distributing honors among the children of the school. It appears in the celebration of national holidays: in the particular events that are celebrated as well as in the historical and contemporary personalities who are chosen to exemplify the qualities of citizenship and worthy community service. It appears in the programs for the general assemblies of the schools: in the various leaders from the community who are brought in to speak to the children. It appears in the way teachers are treated: the amount of freedom and initiative they enjoy, in the extent to which teachers are permitted to take part in the life of their community, and the degree to which the young believe that they are studying under leaders who are more than docile, routine drill-masters in assigned subjects. It appears in the way the community organizes to conduct its schools: in the provision it makes in its schoolgrounds, buildings, and equipment, in the kind of people it chooses to serve on the school board, and in the relation of the members of the board to the administrative and teaching staff. In sum, the moral factor enters whenever and wherever significant decisions have to be made about either the organization, the administration, or the instructional program of the school. All of these decisions, whether they relate to curriculum or to extra-curriculum affairs, exert an influence on the attitudes and the behaviors of the young.

Thus judgments about life values inescapably pervade and undergird the whole process of providing and guiding experience. More than many teachers recognize, a scheme of values—a structure of things considered significant, worthful and right—operates in their endless responses to the daily behavings of their pupils. Many of these educational values concern the very fundamentals of human existence. They have to do with such elemental things as the rights, the responsibilities, the beliefs, the tastes, the appreciations, the faiths and the allegiances of human beings. As we introduce the young to the various aspects of human experience—familial, economic, scientific, technological, political, religious, artistic—we inevitably encourage attitudes and habits of response in and to these affairs. In order to encourage, we must also discourage; in order to foster, we must also hinder; in order to emphasize the significant, we must identify the nonsigificant; and, finally, in order to select and focus attention on certain subject-matters of life, we have to reject and ignore other subject-matters. Were our values different, our selections and our rejections would also be different. The process of selecting and rejecting, of fostering and hindering, of distinguishing the lovely from the unlovely, and of discriminating the important from the unimportant, is unending in education. It is this process of choice and emphasis that defines what is meant by the term *moral* as it is used in this book.

As thus interpreted, the concept of the *moral* refers not primarily to the particular ethical quality of the life interests, outlooks, and practices involved in any given educational program, but rather to the more elemental fact that *choices* among genuine life-alternatives are inescapably involved in the construction and the actual conduct of each and every educational program. These choices necessarily have consequences in the lives of the young, and through them in the life of their society. Viewed from this perspective, education undoubtedly ranks as one of the outstanding moral undertakings of the human race.

11. Value Judgments

E. R. Emmet

* * *

Taste statements and fact judgments lie at extreme ends of the scale. We come now to consider a whole class of expressions of opinion which lie between the two, or about which a vital and interesting thing to be decided is to which of the two they belong. Such statements may be broadly classed as value judgments. We make a value judgment whenever we give a mark or a rating to one thing in comparison with another, whenever we say that A is better than B using 'better' in its broadest, loosest sense. When a schoolmaster gives 35 marks out of 50 to one essay and 29 to another he is making what would ordinarily be called a value judgment. The interesting question that arises however from our discussion in this chapter so far, is whether what he is doing is more closely akin to what one is doing when one says that this stick is longer than that, or to what is being done when one says that this chair is more comfortable than that. Is it a judgment about a matter of fact (in which case it is perfectly possible for it to be an incorrect one) or is it just the expression of a personal preference? Is the judgment objective or subjective?

It is probable that few people would wish to hold either extreme position about this kind of judgment, this act of assessment. Can it really be maintained that the comparative merit to be attached to the essays of, say, a class of 20 is a matter of fact to which there is a precise correct answer, laid up as it were in heaven, if only we can find it? It is worth noticing here that there are two difficulties involved—first the *order* of merit of the different essays and secondly by *how much* one is better than the other. Would it be possible to maintain not only that A's is better than B's and B's better than C's but also that the difference between A's and B's is, say, exactly 2.5 times the difference between B's and C's? (This two-fold difficulty is of course implicit in all scales of preferences or evaluations. To place them simply in an order of merit is to use an *Ordinal* scale; to maintain that it is possible to evaluate the relative amounts by which some are to be preferred to others is to believe in the possibility of a *Cardinal* scale.)

Reprinted with permission of the publisher from *Learning to Philosophise* (New York: Philosophical Library, Inc., 1965), pp. 103–118.

But on the other hand if one discards this view as repugnant to common sense, is one therefore to say that it is simply a matter of taste? That the schoolmaster is arranging the essays in the order which he personally finds pleasing just as he might, if asked, arrange in the order which he finds pleasing a number of dishes? In this case is one bound to say (as one would certainly say as far as the dishes were concerned) that there is no question of his being right or wrong, that any one person's opinion is as good as any other's, or rather that there is no sense in which one can talk about anyone's opinion being 'good'?

Our natural reaction is to say that the truth lies somewhere in between these two extremes. It might also be a natural reaction to object to the last sentences of the preceding paragraph, to say that in a matter of this kind one person's opinion is *not* as good as any other person's, that there is in fact such a thing as an expert opinion.

We must now investigate more closely how, if at all, it is possible to take up a position between the view that marking an essay is a judgment of matters of fact and the view that it is simply an expression of personal preference: we must also examine the concept of expert opinion in this context.

Before we return to the schoolmaster and his essays it will be useful to bring out a few points by considering some other value judgments. Suppose we say 'A is better at chess than B'. Chess is a game and the object is to win. The test therefore for deciding whether A is better than B is the factual, objective one of A playing chess against B—preferably several times—and seeing who wins. If A wins on each occasion easily (i.e. quickly) we should feel justified in saying that the statement is true and that it is a matter of fact. Suppose however that someone who has watched all these games denies this proposition and asserts that in spite of all his defeats B is *really* a better chess player. What should we say to him? It would clearly be natural and fair to ask him what he means by B being better. Perhaps he might say that B is potentially a better player. If A is middle-aged and B a small boy it might be felt that A has won his victories because of his greater experience, but that B has displayed in his style of play a superior grasp of the essential principles which may well make him a very strong player one day. Any sensible discussion about the matter would then be focused on the question whether by A being 'better than' B at chess we are going to mean A having a recent record of victories over B or whether we are going to use the phrase to refer to the probability of B being able to defeat A consistently at some unspecified date in the future. In fact, without defining the matter very accurately, we may well use the phrase to mean some rather vague combination of both; partly a matter of fact, who has won more games of chess; and partly a matter of prediction, who is likely to win more games of chess in the future. Most people in making the judgment as to who is the better chess player are likely to attach considerably more weight to the first of these considerations. It is possible that someone who is making a judgment about two chess players might also consider their styles. B might be a lively, attacking enterprising player while A's style is dull, stodgy, safe. Even though A won more often it might still be maintained by some people that B was the 'better' player. It would be reasonable in such a case to point out that the criterion for being a good chess player must surely be to win or to be likely to win games of chess and that, unless it is claimed that this dashing style may make success in the future more likely, the word 'better' is being used in an odd and irrational sense.

Chess is one of the games in which it is least likely that a consideration of style will form part of the criteria of merit. It is much more likely to happen with a game like cricket when we are considering the rival merits of two batsmen. It would be reasonable to suppose that the criterion for being a 'good' batsman is to score runs or at least to stay in, and that the better batsman will be the one who scores more runs against the same, or approximately equally 'good', bowling; we might also take into consideration the rate at which those runs are scored.

It is common practice however to include also in one's consideration the 'style' or the method in which the runs are made. A good style for a batsman should presumably mean those methods of batting which make it more likely or are thought to make it more likely that he will remain at the wicket and score runs. But there may also enter into it the question of taste; I call A a better batsman because I find his style more pleasing, attractive. It may be argued that if this more attractive style is not actually or potentially the source of more runs I am *wrong* in so calling him. It is nevertheless the case that different people will use the phrase 'good batsman' in slightly different ways, and it is important to notice that what is being done is to some extent reporting on matters of fact, to some extent making a prediction about the future and to some extent expressing a taste preference. 'Better' is inevitably a very vague and loose word in this sort of context and it is always an aid to clear thinking (though sometimes it may be unnecessarily long-winded) if one specifies in what respect better ('more pleasing to watch, though he doesn't make so many runs') instead of using vague terms of commendation.

One more example before we return to our schoolmaster and his essays. Suppose we want to decide whether A or B is the better runner. A can beat B comfortably and consistently over short distances, say less than half a mile; and B can beat A equally comfortably and consistently over any distance more than a mile. Which is the better runner? Obviously the criterion for being a 'good' runner is to run fast. Questions of style may come into it, but again, as with the batsman at cricket, the essentials of a good style at running will be based on principles of efficiency, principles which, if observed, will help one to run more rapidly. The facts in our example are that A is better than B over some distances, B is better than A over others (presumably one could find some intermediate distance over which the result of a race between them would be a dead-heat). If asked to say which is the better, one is being asked to make a comparative assessment of the amount by which A is better than B in one respect and the amount by which B is better than A in another. This is obviously a very difficult thing to do. One could work out a system of marking based on comparative times at different distances and thus define what it is to be a better runner, but any such system would inevitably have an element of arbitrariness about it, and in the last resort it is a matter of opinion, of taste. There is certainly no correct answer. It may well be thought (and I hope that readers are thinking this) that it is a silly question. Why can't we just state what the facts are, who beats whom at what distances? Why do we have to have a comprehensive, overall evaluation?

(An answer might very likely be that there is a cup to be awarded to the 'best' runner, so that a decision or choice has to be made.)

This is a simple version of a process which happens a great deal, though it is a process which may to some extent be wrapped up and disguised. It is not merely the difficulty of the cardinal scale of values to which we referred earlier—the difficulty of comparing the amount by which we prefer A to B to the amount by which we prefer B to C, when A, B and C are of the same kind (e.g. dishes to eat) and the criteria for our preference are of the same sort (tasting nice). It is the difficulty of comparing the amount by which A is superior to B in one respect to the amount by which X is superior to Y in another. (Do you like strawberries better than blackberries by as much as you prefer Ermyntrude to Clarissa?) Obviously if one *has* to give an answer to this question—if one's life depends on it—one will give it, and in fact people are continually making value judgments of this kind. The answer that one gives is not verifiable or falsifiable and certainly it may sometimes be felt that one can answer with conviction and confidence.

(Suppose, for example, that one is passionately fond of strawberries, heartily dislikes blackberries and has only just haltingly, waveringly and without much assurance come to the point of feeling a preference for Ermyntrude rather than Clarissa.)

On the whole most people are likely to feel that the question is a silly one; they might rather resent being forced to give an answer to it, and might feel that any answer they gave would just be the result of a mental toss-up. (How does one set about answering a question like that? Through what mental process is one supposed to go?) And it might well be felt also that to answer a question like that could not conceivably serve any useful purpose.

Answers to similar questions however are implicit in a great many of the evaluations which we do in fact make.

Our diversion from the schoolmaster and his essays was designed to make, by considering simpler cases, three main points:

1. The criteria for being 'better' depend on the purposes of what are being considered. If people are being compared in some particular capacity, for example as chess-players or batsmen, we mean by this the purposes which they have in mind, to win games of chess or to make runs. If physical objects are being compared, for example if we ask which is the 'better' of two houses, we mean the purposes for which they are to be used, to house two people in a cold climate or ten people in a warm one, or whatever it may be.
2. Some of the criteria may be matters of fact in the present, some may be concerned with predictions about what is going to happen in the future, and some may be matters of taste. It is unlikely to be the case that all those who make value judgments about the same things will use precisely the same criteria, though they will generally be at least similar.
3. As we saw in the example of the runners some value judgements will entail a process of comparing the importance or the value of differences between things in one respect, which may itself be a matter of fact or a matter of taste, with differences between the same or other things in another respect. P is a better batsman than Q, but Q is a better bowler. Which is the better cricketer? X is better than Y at French and History, but Y is better at Latin and Mathematics. Which is the cleverer? In some cases in a particular context an act of choice may be necessary: shall I select P or Q for my team tomorrow? But this depends on other circumstances, (e.g. the comparative batting and bowling strength of the rest of the team) and need not imply a generalised value judgment. The number of respects in which a comparison is being made obviously need not be limited to two or three or four.

The process we have been describing, the portmanteau value judgment, 'better than', to cover many different comparisons which may themselves be rather suspect, might be variously described as difficult, impossible or unreal.

We return now to the schoolmaster and his essays. What he is doing is to make comparative value judgments which purport to be precise. He is guided to some extent in making these judgments by matters of fact, whether for example the spelling conforms to the precise rules which exist, whether the punctuation and grammar are correct, though for these the rules are considerably less precise. He is also assessing, perhaps, the extent to which the essays have included certain basic bits of information, certain matters of fact which were parts of what he, the setter of the essay, wanted, the extent to which the writer shows himself well informed about those matters of which his knowledge is being tested. He is also probably assessing the extent to which the writer has succeeded in communicating his ideas clearly, interestingly and entertainingly, and in doing this the schoolmaster will be thinking not only whether he himself is interested and entertained, but also whether others of similar intelligence and education would be likely to be so too. He will be assessing the extent to which the writer has succeeded in achieving the purposes which were set before him.

Some of these assessments are matters of fact, some are matters of taste (whether the reader finds it interesting) some are judgments or predictions about the tastes of other people. What then has to be done is to combine these assessments of different things into an overall value judgment to which a precise numerical value will be attached. The comparative heinousness of a mistake in spelling, an error in punctuation, getting a fact wrong, and being dull, has to be decided. Such comparison is clearly a matter of opinion or taste to which, unlike some of the other assessments that are being made, the notion of expert opinion (to which we shall return later) is hardly applicable. (It is perhaps worth noticing that the schoolmaster may often use marks not simply to measure the gravity of an offence, but as an incentive or deterrent. Brown really must pay more attention to his spelling and to writing legibly, and to encourage him to do so a disproportionately large number of marks may be taken off for failures in these respects.)

It might be said that this operation, this comparison of the size of value differences of different kinds ending up with a result that seems by implication to claim to be precise, is so absurd, is so obviously (especially to anyone who has tried to do it) not precise, that one wonders why it is done.

It is of course the result of a tendency, which seems to become more marked as the world becomes more competitive, to want to put things and people in an order of merit. It is partly because it has come to be thought that since with the advance of science man has increasingly succeeded in measuring things, everything (intelligence, literary skill, etc.) is measurable, and there has grown up the very dangerous tendency to try to be precise or to claim to be precise when no precision is possible, and to suppose, when there is a failure to achieve precision, that this is due to some inadequacy in ourselves or in our tools or methods of measurement, instead of being due, as is very often the case, to the nature of what is being considered.

Only by drawing up a clear-cut precise scale of values or marks for different points could it be said that one essay was worth 35 marks and another 29, but the drawing up of this scale of values would be to introduce an artificial precision concerning what is bound in the last analysis to be a matter of opinion or taste.

Nevertheless, although the precision implicit in the numerical evaluation of essays may be generally speaking unjustified and artificial, it is certainly and obviously not the case that the operation of marking essays is without value. As we have pointed out the writer is trying to achieve certain purposes (for example to be intelligible and interesting) and the schoolmaster is assessing the extent to which these purposes have been achieved as far as his reactions are concerned and are likely to be achieved with regard to the reactions of other people. Experiments are often made to determine the extent to which different examiners agree about the comparative merits of essays and, as might be expected, although agreement on the whole is not very close, it is usually close enough to provide some justification for the operation of marking. In other words there is likely to be a consensus of opinion that certain essays are more intelligible and interesting than others even though it is not possible to measure by how much. If however the assessment of some few examiners is markedly different from that of the majority, this may well be thought to show how little the few know about it, that they are not experts; and it might be maintained that the more 'expert' the opinion the greater the measure of agreement is likely to be.

The time has come to examine rather more closely the concept of 'expert opinion', to see what part, if any, it has to play in affairs which are matters of taste.

EXPERT OPINION

Let us take a simple case first, the comparative 'comfortableness' of a number of chairs; a case where the comparison would be generally agreed to be subjective and not a matter of fact. Is there any sense in which it would be reasonable to talk about expert opinion in this context?

What the expert can do is to study the human frame and the feeling-reactions of himself and other people to chairs of various shapes and designs, and to make deductions and generalisations about what people do find comfortable and are likely to find comfortable and why. The advice of such an expert will be useful to the manufacturers of the chairs and also to individuals who are selecting chairs. 'This chair here, for example, you may not find very comfortable when you first sit in it, but my experience and that of other people has been that if you try it for a bit and get used to it you will eventually find it more comfortable than anything you have ever sat in; you will get a higher quality of comfort-pleasure than you have ever received.'

What the expert does is to advise people how their tastes may be best satisfied, how the purpose of giving or getting comfort-pleasure is most likely to be achieved. And he becomes an expert by studying the subject matter.

Similarly an expert in food may advise people what they are likely to enjoy tasting, and make recommendations whereby they may increase the pleasure of eating. Just as, since people are constructed in roughly the same way, there are certain general principles derived from experience as to what they are likely to find most comfortable, so since people's palates and taste buds are similarly constructed there are certain general principles as to what they are most likely to enjoy eating. But these general principles are clearly not precise, there will be many exceptions, they will vary from one community to another, and there will certainly be elements of habit and convention in the formation of these principles.

Notice that the expert may make statements about what people are likely to find comfortable and what they are likely to find nice and these statements may be verified or falsified by events (and the verification of his statements will also serve to some extent as a verification of his expert status). It would however be misleading if he were to make statements about what *is* comfortable, what *is* nice; if his statements instead of appearing to be predictions or recommendations, were to be interpreted as telling people what they *ought* to find comfortable to sit in or nice to eat.

In fact this is not at all likely to happen with comfort, and not very likely to happen with food. It is rather more likely however to happen with wine. The expert in this case becomes the 'connoisseur'. He will be knowledgeable and an expert not only in the sense that he is likely to be able to tell the source and perhaps even the date of a wine by tasting it but also that what he says about which wines are good will be accepted as authoritative and right. When he says that one wine is 'better than' another, however, it may sometimes be thought that he is doing something rather more than, and different from, making a prediction about which wine people in general will prefer. But obviously wines are for drinking and it can hardly make sense to use any criterion for judging the comparative merits of wines other than the extent to which people enjoy drinking them, though one might certainly include under 'enjoyment' here, not only the experience at the time but also the after-effects, pleasant or otherwise, any curative or toxic properties that the wine might have. The connoisseur however might make the claim that even though the majority of people would prefer A to B, nevertheless B is better than A. What could be meant by such a claim and can it be substantiated?

Such a claim could clearly only be based on the pleasure that *someone* derives from drinking it. He might say that the pleasure that the expert derives from B is of a higher quality than that derived from A, and that if other people had the opportunities and would take the trouble to train their palates they too would appreciate this. He is making the recommendation that potentially the pleasure derivable from B is to be preferred or would in general be preferred to that derived from A. This recommendation or prediction is in principle verifiable and would seem to provide a reasonable basis for the expert to say that B is better than A.

Let us see now how the concept of expert opinion applies to the schoolboy's essays or rather to the field of literature in general.

EXPERT OPINION AND LITERATURE

The purposes of literature are considerably less clear-cut and more various than the purposes of the chair, the food or the wine. They may be to inform, to explain, to communicate ideas, to stimulate new ways of looking at things, to amuse, to entertain, to persuade, to rouse emotions, to communicate deep emotional experiences or to provide them. It has been said that 'the study of it [the literary critic's literature] is, or should be, an intimate study of the complexities, potentialities and essential conditions of human nature.' (F. R. LEAVIS, *The Common Pursuit*, p. 184.) The purposes of the schoolboy's essay must certainly be included among the general purposes of literature, it should be an exercise in achieving one or more of these purposes, and its 'value' should clearly depend on the extent to which it does these things or is thought to be likely to do these things. All these purposes, it should be noticed, are expressed in terms of the effects which the literature has on those who read it; literature is for reading as wine is for drinking. But though drinking is for pleasure, reading need not be.

Clearly the 'expert' about literature—the literary critic, the schoolmaster the university don—in expressing an opinion about the merits of a certain work, will be partly making an assessment about matters of fact—whether what is written is true, accurate, properly spelt and punctuated and grammatically expressed (though it is only in certain departments of literature that these things will necessarily be counted as merits); partly assessing the extent to which the work does to him what it is intended to do—entertain, delight, interest, inform, instruct, emotionally move or generally improve—and partly making a prediction about the extent to which it is likely to do these things to other people, perhaps especially in some cases a particular class of other people, those who are also 'experts' in the relevant field.

Literary critics would be the first to express relief at the fact that on the whole they are not expected when writing their criticisms or opinions to attach a precise mark to various works as the schoolmaster is often impelled to do to his essay, though they may sometimes attach 'ratings' they may sometimes produce a list of the ten 'best' books of the year or of the century, and they will almost certainly attach to the books they review adjectives, usually rather vague ones, of approval or disapproval. To what extent then, if any, can it be said that literary merit is a matter of fact, objective, and to what extent a matter of taste, subjective?

There is no doubt that the experts, the literary critics, do often write as though literary merit was something objective, as though there was a sense in which it could be said that some books were 'better' than others and that anybody who thought otherwise was just wrong. They would be likely to maintain stoutly that majority opinion is not what determines whether a book is 'good' or not, and it is often implied that, though its merit is obviously much more likely to be discovered by expert opinion, there is a sense in which that merit would still be 'there' whether the experts discovered it or not.

Those who disagree with the concept of literary merit as being something objective, valid for all men, would maintain that all that the critic can usefully and justifiably do by way of assessment is to report on the effect that the work has on him, perhaps to suggest reasons why this effect is as it is, and explicitly or implicitly to make a prediction about the effect it is likely to have on other people. It will sometimes be claimed that these effects are capable of being exceedingly important; the study of literature may be regarded 'not only as a discipline in itself, but as a social and moral force, at once a prophylactic and a remedy for the corruption and degradation inherent in our materialistic society' (J. B. Bamborough. *The Spectator*, 25 October 1963, describing a conviction attributed to F. R. Leavis.) But the importance of the effect makes no difference to the argument that it will vary from person to person, and that the value will not be valid for all men.

In admitting the fact that the experts hold strong views about their scales of preferences and sometimes, but by no means always, agree among themselves on a preference that is contrary to majority preference, the subjectivists might introduce the idea of a higher quality of pleasure and a progression, as a result of expert knowledge and experience, from a lower quality to a higher. It is a matter of fact for example that the writings of Jane Austen will generally be described as 'better than' those of, say, Edgar Wallace, in spite of the probability that they have been read and enjoyed by many fewer people.

An objectivist would say that this is *because* they possess greater literary merit, (or that this amounts to the same thing as saying that they possess greater literary merit), and that anyone who thinks otherwise is lacking in literary appreciation and is just wrong. A subjectivist would reply that this is a misleading way of describing the facts. One can say that some people prefer Jane Austen, others Edgar Wallace, and that though there can be no question of either group being right or wrong, it is nevertheless a fact that those who study the matter closely and do a lot of reading find that the pleasure or general benefit they get from Jane Austen seems to them to be of a higher quality; many cases are found of people progressing, as they would describe it, from Wallace to Austen, but not many cases of people making a similar change of preference from Austen to Wallace. (It is possible of course that this description as 'progression' may itself be influenced by what would be a widely held view that one 'ought' to prefer Austen to Wallace.)

Those who have made such a progression feel entitled therefore to recommend that others should do likewise. They point to the superior intellectual pleasures that may be derived from the study of 'good' literature in the same sort of way as the expert on food and drink may draw attention to the superior tasting-pleasures that may be derived from the cultivation of the palate and the study of 'good' cooking. The subjectivist might make the further point that he would not want to express an overall preference for one author over another. It all depends—on one's mood, on whether one wants to be relaxed or stimulated and so on.

'But yet', the objectivists might reply, 'everyone knows that some books *are* better than others; this alternative way of describing it can't disguise the fact that either a book has literary merit or not. And once this is admitted to be a matter of opinion, there are no standards, everything is shifting and chaos reigns.' Objectivists also tend to feel that the subjectivist view degrades the importance of literature and the intellectual and aesthetic delights that can be derived from it.

To the first objection the subjectivist would reply that there are certainly no absolute standards in the matter, and that 'literary merit' as it seems to be thought about by his opponent is a myth; he is thinking about it in the wrong way and the objection begs the question. He would go on to say that the absence of absolute standards, valid for all men, and the emphasis on opinion or taste need in no way lessen the pleasures, perhaps the improvement, to be derived from literature, or alter the fact that such experiences are usually regarded by those who have had them as preferable to the pleasures of the table.

But these and similar objections are raised even more strongly against the subjectivist view of value judgments of beauty, and as the arguments on either side are obviously very similar for beauty and for literary merit it will be useful to say something now about beauty in general. . . .

12. Justifying Curriculum Decisions

Israel Scheffler

Decisions that confront educators are notoriously varied, complex, and far-reaching in importance, but none outweighs in difficulty or significance those decisions governing selection of content. In view of recent talk of "teaching children rather than subject matter," it is perhaps worth recalling that teaching is a triadic relation, describable by the form "*A* teaches *B* to *C*," where "*B*" names some content, disposition, skill, or subject. If it is true that no one teaches anything unless he teaches it to someone, it is no less true that no one teaches anybody unless he teaches him something.

We do not, moreover, consider it a matter of indifference or whim just what the educator chooses to teach. Some selections we judge better than others; some we deem positively intolerable. Nor are we content to discuss issues of selection as if they hinged on personal taste alone. We try to convince others; we present ordered arguments; we appeal to custom and principle; we point to relevant consequences and implicit commitments. In short, we consider decisions on educational content to be responsible or justifiable acts with public significance.

If these decisions are at once inescapable, important, and subject to rational critique, it is of interest to try to clarify the process of such critique, to state the rules we take to govern the justifying of curricular decisions. Such clarification is not to be confused with an attempt to justify this or that decision; rather, the aim is to make the grounds of decision explicit. Furthermore, clarification cannot be accomplished once and for all time but is rather to be seen as a continuing accompaniment to educational practice.

It is the task of clarification that I shall consider here. I shall offer an analysis of the process of justification along with suggestions for justifying decisions on curriculum.

Reprinted with permission of the author and publisher from *School Review* 66 (Winter 1958): 461–472.

What is subject to justification? A child may be asked to justify his tardiness, but he would never be asked to justify his cephalic index. Fiscal policies and choices of career are subject to justification, but typhoons and mountain ranges are not. Justifiability applies, it seems, only to controllable acts, or *moves,* as they will henceforth be called.

In this respect, justifiability is paralleled by the notion of responsibility, with which indeed it is intimately related. If am held responsible for violating a traffic regulation, I expect to be subject to the demand that I justify my violation. Conversely, the child who is called on to justify his late arrival for dinner is being held responsible for his tardiness. The child may escape the need to justify his lateness by denying his responsibility for it. He can deny his responsibility by denying that his lateness was a move at all, by claiming that it could not be helped, was not deliberate or subject to his control.

Now that I have asserted that only moves are justifiable, I must immediately add one qualification. In ordinary discourse, we do not limit justifiability to moves. A city-planning group may debate the justifiability of a projected highway. However, the issue here can ultimately be construed as the justification of moves calculated to produce the highway in question. In general, ostensible reference to the justifiability of non-moves may be construed as a shorthand reference to the justifiability of moves appropriately related to non-moves. Where such moves are lacking, the justification of non-moves fails to arise as an issue. Thus, while we may speak of highways and courses of study as justifiable, we do not inquire into the justification of comets or rainbows. Justifiability may, then, be taken as a universal property of moves; and those that are, in fact, justified comprise a subclass of moves with a certain authority in our conduct.

How are moves justified? If the justified moves represent a subclass of all moves, then to justify a particular move requires that we show it to be a member of this subclass. If no further specification of this subclass is given, we have a relative sense of justification.

Consider chess: we have a board and the standard pieces. We understand what constitutes a move, and we have rules that permit only certain moves. These rules, in effect, define a subclass of all moves. For a player to justify his move as a chess move requires that he show that it belongs to the chess subclass. Such justification is strictly limited, for it depends clearly on the set of rules that define chess. There are an indefinite number of other rule-sets singling out alternative subclasses of moves. A move justified for chess may not be justified for checkers and vice versa. A chess player justifying his move is not implying that chess is superior to checkers. He is only showing that his move conforms to the rules of chess. Hence we cannot speak, strictly, of a move on the board as justified in general or in the abstract; we have to specify also the operative rules.

Some processes of justification resemble the justification of moves in chess and in other formal games. These processes have a well-specified set of rules defining appropriate moves. Justification consists of showing that a move conforms to these rules, that is, belongs to the subclass singled out by them. There is no thought of justifying the set itself as against alternatives. Though it may not be explicitly stated, it is evident that moves are being justified only relative to this set. These conditions seem to apply when, for example, we consider Smith's driving on the right side of the road (in Massachusetts) to be justified. Driving on the right conforms to Massachusetts traffic rules. We are by no means claiming that these rules are unique or superior to alternative rules, for example, rules of countries where driving on the left is prescribed. What is involved here is relative justification. Traffic regulations are, in an important sense, like chess rules or games in general. For one reason or another, we may be interested for the moment in playing a certain game or in seeing what the game demands in a particular case. But the existence of alternative games fails to upset us, nor is the comparative justification of the games as such in question.

Relative justification is not limited to such clear cases as traffic control. Much of our conduct falls within the range of less well-defined rules, or social practices and traditions. Much of the time, we justify this conduct simply by appeal to conformity with established practice. Nor should it be supposed that such justification is always as uncomplicated as that of our traffic illustration. Often a move is justified by appeal to a rule, and that rule by appeal to another. For example, Smith's right-hand driving may be justified by a demonstration of its conformity with Massachusetts law and this particular law by conformity with traditional legal practice throughout the United States. Though various levels are distinguishable, it is still true that the justification as a whole is here carried out in relation to American practice. That is, such practice sanctions a class of certain subclasses of moves, one subclass of which includes the move in question. In effect, one "game" is justified by showing its imbeddedness in another, larger "game."

The relative sense of justification is, however, not exhaustive of the types of justification that one used and, in itself, is hardly satisfactory for many purposes since every move is both justified and unjustified, in relation to appropriately chosen sets of rules. If I am not, as in a game, asking what move I ought to make in order to comply with some particular set of rules, but am asking what move I ought to make at all, the relative sense of justification will be of no help whatever. At best, it can lead me to another query of the same sort on a new level and leave me equally undecided there. The non-relative, or general, request for justification is, furthermore, one we often make or imply, and in the most important departments of life—belief, social relations, individual choices.

When we decide broad educational issues, we are often asking not merely what jibes with American practice, past or present, but what is generally justified, whether or not it is sanctioned by practice. The desire to evade this general question is understandable because it is difficult. But this evasion, I think, is responsible for much of the inadequacy of value-discussions in education. Two tendencies seem to develop. A move is defended on grounds of its conformity with American practice, and the question of the justification of this practice itself is not considered at all. Or it is flatly asserted that it is the duty of the teacher to conform to the educational practices of his society, an assertion which, besides calling on a non-relative notion of duty that is itself uncriticized, seems to many schoolmen to be far from obvious.

Both the nature of this general request for justification of acts or moves and the possibilities for dealing with it may be illuminated by comparison with belief. To know that a belief is justified in relation to certain evidence does not provide general justification unless we have confidence in the evidence to begin with. With this initial confidence or credibility, we can proceed to provide ground for our belief. Roughly speaking, what we seem to do is to justify beliefs that not only hang together logically but also, as a family, preserve this initial credibility to the highest degree. We judge belief in question by its general impact on all other beliefs we have some confidence in. No matter how confident we are of a particular belief, we may decide to give it up if it conflicts with enough other beliefs in which we have a higher degree of confidence.

In practical situations, of course, we do not actually take all our beliefs into account. We concern ourselves, rather, with a limited domain of beliefs that we feel are interdependent. Furthermore, we do not make piecemeal estimates of the impact of each belief on the credibility of the mass of our beliefs in this domain. Instead, we use summary rules of varying generality. These rules are quite different from the rules of chess, however. They are not simply chosen at will but mirror, in a systematic and manageable form, our confidence in particular beliefs, classes of beliefs, and combinations of beliefs. Theoretically, there is no control (except perhaps that of the demand for consistency) over the design of games, no external requirements they need meet. The rules we use in general justification of belief are subject to the requirement that they be true to our credibilities on the whole. If a rule conflicts with our credibilities, it will be scrapped. We may say that rules are justified if they adequately reflect our credibilities by selecting those groups of beliefs that rank highest in this regard. A particular belief, then, is justified by its conformity with rules so justified. In effect, it is justified if it hangs together with that family of beliefs that as a whole commands our highest degree of confidence.

Formal logic as a code of valid inference provides an instructive example. People judged good and bad arguments long before the Aristotelian code. The latter was intended to systematize individual judgments and derives its authority from its adequacy as a systematization. When we now refer the justification of a particular inference to the ruling of Aristotle, our procedure depends on our confidence in this adequacy. It is a shorthand way of seeing whether or not the inference belongs with the mass of inferences we find most acceptable. Theoretically, no element in our procedure is free from future reappraisal. If, at some future time, we find that the existing code demands the abandonment of an inference that we value or the acceptance of an inference that we detest, we may alter the code. If an inference we are attached to conflicts with our code, we may give up the inference. There is a mutual adjustment of rules and instances toward selection of that family of instances that, as a whole, has the highest claim on our acceptance. An instance or rule that interferes with such selection is subject to rejection.

Codes of deductive or inductive logic may be construed as definitions of valid inference, not in the sense in which definitions may be used to introduce coined terms, but rather in the sense in which we set about defining a term already in common use, where this use controls our definition. The man who invented Scrabble was defining the game in the first sense by laying down rules that were labeled "Scrabble Rules." On the other hand, if a man from Mars were to arrive in the midst of a Scrabble tournament, without benefit of prior study of the official rules, and were asked after some hours to define the game, his task would be considerably different from that of the inventor. He would have to observe, guess, and test, to determine whether his proposed list of defining rules actually squared with the moves of the players. He would be attempting a definition in the second sense.

Even this task would be simpler than that of defining valid inference or, indeed, of defining any term in general use. For our man from Mars could always, as a final resort, check his definition against the official rule book. But for valid inference as for other notions in general use there is no official rule book at all. We start by proposing a definition that will serve as a simplified guide to usage but continue to check our proposal against actual use. We justify a particular use of a term by appeal, not just to any definition, but to one that we feel is itself justified by adequate codification of usage. In effect, we justify a particular use by checking it, through adequate definitions, against all our other uses.

These examples illustrate what we may expect and what we may hope to accomplish in the general justification of moves. Justification in relation to a given set of rules is useless unless the latter are themselves justified. But further relative justification by reference to other sets of rules is fruitless. Somewhere there must be control of rule-sets by initial commitments to moves themselves. The rules we appeal to in justifying social moves are rules that we hope are themselves adequate codifications of our initial commitments. The rules we appeal to select those families of moves that, as wholes, command our acceptance to the highest degree. Without initial commitments there can be no general justification, any more than there can be real or controlled definition without initial usage. But the fact that we are attached to a particular move does not mean that we cannot check it against all others we are committed to (by way of rules), any more than our attachment to a particular locution means that we cannot check it against others we hold proper (by way of controlled definitions). Our legal and moral rules serve, indeed, to guide the making of particular moves, but their guidance depends on their presumed adequacy in codifying our initial commitments to moves, on the whole.

In accordance with the two senses of justification just discussed, we may distinguish two levels of justification of educational decisions. On one level, justification involves conformity with a set of rules, reference to which may be implicitly understood. Here the issue is relative. We ask, "Is such and such decision justified according to rule-set *S*?" For many purposes, the question is legitimate and important, but the answer is often far from simple, even when the rules are fairly well defined. Relative justification is often a highly complicated, intellectually engaging business. To appreciate this fact, one need only recall that there is a whole profession (law) devoted to solving just such questions as the conformity of cases to rule. In education, such justification seems to relate not to specific laws but to broad social practices and traditions, the formulation of which has to be abstracted from our history and is itself a difficult job. Still, such traditions are often cited and used as a lever for changing laws as well as individual decisions.

Yet, legitimate as relative questions are, they do not exhaust our queries in educational contexts. We are not always interested merely in knowing that an educational move conforms to some code. We want to press the issue of deciding among codes. We ask that our moves be justified in terms of some justified code. If our previous analysis is correct, we are seeking justification by rules themselves controlled by the mass of our initial commitments. Of the two levels of justification in educational contexts, the relative type is familiar. The practical issues here may be complicated, and one factor often adds to the complexity: ostensible questions of relative conformity to a given rule may be decided, partly at least, on independent moral grounds. Yet, many of these issues seem familiar in outline. The understanding of general justification presents a more formidable task, since the formulation of relevant rules is of the difficult variety illustrated by the attempt of the man from Mars to codify the rules of a game by watching the play. We need to do something of this sort, but far more complex, since the activity involved is our own and touches on our fundamental commitments.

What rules do we appeal to in general justification of educational decisions on content? The answer to this question consists of a set of rules, not assertions, but the process of compiling an adequate set of rules is as empirical a task as can be imagined. Definitions are not assertions; but to compile a set of definitions one needs to call on all sorts of information, hypotheses, hunches—and the resulting set is always subject to recall, if not to falsification. It is with such qualifications that I offer my list of rules relating to decisions on curriculum. This list should be construed as a hypothesis, tentatively offered and inviting criticism. If it proves wrong, the process of correcting it will itself help clarify the grounds of our curricular decisions.

To simplify our considerations, let us avoid, at least at the outset, the problem of formulating special, complicated rules for deciding on content to be taught at a particular time and in particular circumstances. Let us consider instead all the content to be learned by a child during his formal schooling. Without worrying, for the moment, about the functions of particular segments of this content, let us ask instead what we expect of the content as a whole. Let us, further, state our rules in terms broad enough to allow for practical judgment in applying them to cases.

The guiding principle underlying the following rules is that educational content is to help the learner attain maximum self-sufficiency as economically as possible.

Presumably, self-sufficiency can be brought about economically or extravagantly; context should be selected that is judged most economical. Three types of economy are relevant. First, content should be economical of teaching effort and resources. Second, content should be economical of learners' effort. If a very strenuous way and a very easy way of learning something are otherwise equal, this rule would have us select the easier course. Some such principle seems to figure often in educational discussion. For example, the linking of subject matter to children's interests is often defended on grounds that this technique facilitates learning, and even opponents of this approach do not argue that these grounds are irrelevant. It is important, however, to specify that our rules all contain a tacit clause: "other things being equal." It may be argued, for example, that the strenuous course makes for perseverence and other desirable habits, as the easy course does not. Here, however, other things are not equal, and the present rule fails to apply. Criticism of extremism in progressive education, for instance, may be interpreted as insisting that the "interest" principle never stands alone but is always qualified by the clause "other things being equal." Once qualified, the rule stands, in my opinion. There is no positive virtue in unnecessarily taxing the learner; his energy may better be saved for other tasks.

Finally, we must consider economy of subject matter; content should have maximum generalizability or transfer value. The notion of generalizability is, however, ambiguous. Accordingly, two types of subject-matter economy need to be distinguished. First, is there an empirically ascertainable tendency for the learning of some content to facilitate other learning? Presumably, this sort of question was at issue in the controversy over classics, and it was discussed in terms of empirical studies. Second, is the content sufficiently central logically to apply to a wide range of problems? This is not a psychological question but one that concerns the structure of available knowledge. Nevertheless, it is through some such principle of economy, in the logical sense, that we decide to teach physics rather than meteorology, for instance, where other considerations are balanced.

The most economical of content in all the aspects described must still meet the requirements of facilitating maximum self-sufficiency. It should be obvious that we do not necessarily, or ever, apply first the rules of economy and then the rules of self-sufficiency. These rules represent, rather, various requirements put on content, and we may apply them in various orders or simultaneously. We turn now to the rules of self-sufficiency.

Content should enable the learner to make responsible personal and moral decisions. Self-awareness, imaginative weighing of alternative courses of action, understanding of other people's choices and ways of life, decisiveness without rigidity, emancipation from stereotyped ways of thinking and perceiving—all these are bound up with the goal of personal and moral self-sufficiency. The problem of relating school subjects to such traits is an empirical one, but I think it extremely unlikely that a solution is to be found in the mechanical correlation of each subject to some one desired trait. Rather, the individual potentialities of each subject are likely to embrace many desired habits of mind. The use of literature to develop empathy is often noted. But to suppose that this function is restricted to literature is to impoverish our view of the potentialities of other subjects. Anthropology, history, and the other human sciences also offer opportunities to empathize. But even the natural sciences and mathematics may be seen not merely as technical equipment but as rich fields for the exercise of imagination, intuition, criticism, and independent judgment.

The making of responsible personal and moral decisions requires certain traits of character and habits of mind, but such decision-making also requires reliable knowledge, embodied in several areas of study. Psychology, anthropology, and other human studies illumine personal choice; history, political science, economics, sociology and related areas illumine the social background of choices of career and ideology.

We have spoken of personal and moral self-sufficiency, but this is not enough. Since personal and moral decisions are not made in a vacuum, their execution requires technical skills of various sorts. Content should thus provide students with the technical or instrumental prerequisites for carrying out their decisions. What this goal may require in practice will vary from situation to situation; but, speaking generally, mathematics, languages, and the sciences are, I believe, indispensable subjects, while critical ability, personal security, and independent power of judgment in the light of evidence are traits of instrumental value in the pursuit of any ends. In creating curriculums, the notion of technical or instrumental self-sufficiency provides a counterbalance to emphases on the child's interest. For subjects unsupported by student interest may yet have high instrumental value for the students themselves. To avoid teaching them such subjects is, in the long run, to hamper their own future self-sufficiency, no matter what their future aims may be. Thus, it is misleading to label as an imposition of adult values the teaching of instrumentally valuable subjects.

Finally, beyond the power to make and to carry out decisions, self-sufficiency requires intellectual power. Content, that is, should provide theoretical sophistication to whatever degree possible. Here we may distinguish between logical, linguistic, and critical proficiency—the ability to formulate and appraise arguments in various domains, on the one hand, and acquaintance with basic information as well as with different modes of experience and perception, on the other. The danger here, a serious risk of general education programs, is that of superficiality. But ignorance is also a danger. How to avoid both ignorance and superficiality is the basic practical problem. I should hazard the opinion that the solution lies not in rapid survey courses but in the intensive cultivation of a small but significant variety of areas.

Part Five: Concepts of Moral Education

Introduction

The three essays in this section deal in various ways with the importance of cognition and rationality in moral education. The first essay, by Gilbert Ryle, also considers issues that philosophers have been debating since Plato raised them in the dialogues. In the *Protagoras*, for instance, Plato has Socrates ask whether virtue can be taught, and if so, why are there no expert instructors on hand to teach it? As Ryle reminds us, Protagoras suggests to Socrates that teachers of virtue are unnecessary because the acquisition of virtue is similar to the acquisition of language. Later, Aristotle advanced the view that learning virtues is analogous to developing skills or acquiring habits. That is, "we become just by doing just acts, temperate by doing temperate acts, brave by doing brave acts."[1]

Ryle finds neither of these hypotheses entirely satisfactory; nor does he subscribe to a purely cognitivist solution to the problem. Learning virtuous conduct is less a matter of "learning that" or "learning how," he thinks, than of "learning to be," and learning to be is accomplished in large part without benefit of explicit instruction.

In contrast to Ryle, Paul Hirst, in the final chapter of his book, *Moral Education in a Secular Society*, makes "the case for explicit moral education." Hirst also places more emphasis on cognitive and rational analysis in his approach to moral education. "In moral education, as in any other area of education," he writes, "what is asked of the teacher is a total commitment to the development of rational autonomy in both thought and action."

In addition, Professor Hirst argues that moral education should be presented as a separate subject or area of study (though not necessarily labeled "moral education") rather than treated more or less in passing in, say, English. Hirst is convinced that the full time alloted to a distinct subject is required to properly acquaint students with basic moral principles and the complexities of moral discourse. He insists, moreover, that teachers who are assigned courses in this area should be reasonably well versed in moral philosophy.

Other significant points touched on in this discussion include the need to provide students with the factual knowledge that is a prerequisite for the consideration of moral questions, and the importance of practice in dealing with moral issues and problems.

It has been suggested that moral education in the United States is to a large extent reducible to three basic perspectives, which "may be labeled *affectivism, developmentalism,* and *cognitivism.*"[2] The first two writings in this section (though they were not written by Americans) reflect certain aspects of the affectivist and cognitivist approaches, respectively. The third essay includes elements of all three. It claims that a nonrational (not irrational), more or less affective, approach to moral education is unavoidable with younger children who are not yet capable of responding to abstract moral reasoning. With regard to older children, on the other hand, the article recommends an emphasis on rationality and cognition. In other words, the article suggests that we "conceive of moral education as including two fairly well-defined levels or phases," the first being concerned mainly with socialization and the second with critical thinking about moral issues.

Essentially this essay is an attempt to synthesize the thinking of several leading moral and educational philosophers whose writings have contributed significantly to our theoretical understanding of moral education. A follow-up article addressed primarily to educational practitioners attempted to move a bit further in the direction of practical application.[3]

NOTES

1. *Nichomachean Ethics,* II, i, 4.
2. Michael Scriven, "Cognitive Moral Education," *Phi Delta Kappan* 56 (June 1975): 689–694.
3. Peter F. Carbone, Jr., "Teaching Students to Think Critically," *NASSP Bulletin* 59 (January 1975): 64–72.

13. Can Virtue Be Taught?

Gilbert Ryle

Several of Plato's early dialogues raise the question 'Can virtue be taught?' It has not been fully answered yet, so let us look at the question again. One thing that very properly struck Socrates was this. Subjects like arithmetic, geometry, astronomy, rhetoric, wrestling, medicine and so on can be taught, and there exist professional teachers of them. For a person to become good at engineering, swimming or shooting he needs to be and he can be instructed by experts; and they will make him good at it, if he has the aptitude and the industry. But for a person to become a good man, as distinct from a good carpenter, a good navigator or a good astronomer, we do not seem to provide or require expert coaches. There are no technical colleges where honourableness is taught; there are no periods set aside on the school timetable for classes in unselfishness; Oxford and Cambridge colleges offer no scholarships in ungreediness, frankness or humaneness. If virtue were teachable, we should expect there to exist experts to teach it and examiners to test progress in it. But such teachers and examiners do not exist. So it begins to look as if virtue cannot be taught. If a person is honest or considerate, perhaps he is so because he was born to be so, as he was born to be blue-eyed or snub-nosed, or as he is congenitally asthmatic.

Yet notoriously there is a difference between a bad upbringing and a good upbringing, and among what we credit to a good upbringing are things like frankness, command of temper and loyalty. These are good qualities of character, or—nasty word!—virtues, and if we think that they are due partly to upbringing, then, we do not think that they are entirely congenital. No one has learned to be blue-eyed or short-sighted; but most people do seem to have in some degree learned self-control and kindliness. None the less, neither the State nor the university nor the school appoints any specialist instructors from whom these things are learned. What is more, we ourselves, when we consider the matter, find the idea somewhat ridiculous that

Reprinted with permission of the publisher from *Education and the Development of Reason,* eds. R. F. Dearden, P. H. Hirst, and R. S. Peters (London: Routledge & Kegan Paul Ltd., 1972), pp. 434–447.

people should, even in Utopia, be appointed as specialist instructors in industriousness, self-denial and valour. We feel vaguely that there is here nothing specialized for them to be specialists in; nothing that they could be experts in that we do not ourselves, all of us, know quite well without recourse to lectures and textbooks. We do speak of knowing the difference between right and wrong, but we do not think of this knowledge as belonging to a corpus of knowledge which scientists or scholars are forever advancing beyond our laymen's ken, as they are doing with astronomy, philology and mathematics.

In the Platonic dialogue that goes by his name Protagoras tells Socrates that he ought not to be puzzled by this. He adduces an interesting parallel. One of the most important things a child has to learn is to speak and understand his own mother tongue. Yet, although knowledge of Greek is not born in the Greek infant, as his blue eyes and his asthma are born in him, still there are in Greece no professional teachers of the Greek tongue; and this for the simple reason that at the start the child's own mother and father, and afterwards his companions and all the people that he meets, are his unprofessional teachers of conversational Greek. The child picks up his mother tongue perfectly adequately not from anyone in particular, and certainly not from any professional linguists, but from everyone in general. In the same way, Protagoras suggests, though we do indeed learn our standards of conduct, we do not have to learn them in any set lessons conducted by any appointed pundits; we learn them from Everyman in the home, in the streets, in the playground and in the market-place.

Protagoras' solution of the problem is rather an attractive one. 1. We do think that learning standards of conduct is something that takes place when we are pretty young, like learning our mother tongue. 2. We do think that standards of conduct are not esoteric things that only a few specialist researchers know anything about, but are exoteric things that nearly everyone knows nearly everything about. There is no gap here between amateur and expert and so there are no experts and no amateurs either, just as there are in England no experts or amateurs in colloquial English. 3. We are, however, quite ready to allow that just as there are in English occasional niceties and complexities, to master which we really do have to consult dictionaries, textbooks of grammar or phoneticians, so there are occasional moral issues which are so intricate or so unfamiliar that we really do need to take advice from some counsellor of wide experience, lively imagination and shrewd judgment. Soon we shall see that Protagoras' solution does not do the whole trick; but it has cleared the air a bit. It has shown us, by the analogy of learning our mother tongue, how we can continue to maintain that moral standards are not inborn but have to be learned, while allowing that the learning of them does not require the existence of professors of probity, charity and patience.

Now for a different but connected point. One reason, and a bad reason, why the idea strikes us as ridiculous that there should exist expert tutors or crammers in fidelity, modesty or generosity is this. We all know and, unfortunately, are unable to forget that in some teaching, and particularly in much classroom teaching, the teacher is partly occupied in telling his students that this or that is the case. His students have to remember, if not to memorize, the things that he has said, and they have learned their lessons if, afterwards, they can say *viva voce* or write down the things that he told them. This is instructing by dictating. It is, disastrously, this brand of tuition that we all first think of when we think in the abstract about learning, lessons, teachers, etc. We forget, what we all know perfectly well, that there are lots and lots of things that can indeed be learned, and yet cannot be got by heart or regurgitated. To take a simple example. No child can ride a bicycle at first go. He has to learn how to balance, steer, propel and brake the machine. But he cannot learn these things simply by memorizing, however accurately, his father's verbal information about the requisite motions and coordinations. The father cannot just lecture his son into mastery of the bicycle. No, the boy has not just to be told what to do; he has to be shown out on the gravel what to do; he has to be made to do it; he has to try out everything time and time again, try it out in operations on the recalcitrant bicycle itself. He is trained or coached much more than he is told. The skills are inculcated in him by example and by exercise; they are not and could not be crammed into him by mere didactic talking. The same thing goes for cricket, rowing, carpentering, shooting, singing, swimming and flying. It also goes for a lot of more academic or intellectual things like calculating, translating, pronouncing, drawing, dissecting, reasoning and weighing evidence. No one learns to do such things even badly, and much less to do them well, unless, besides being told a few memorizable things, he is also put through lots of critically supervised exercises in the operations themselves. Acquisition of skills and competences comes, if at all, only with practice.

Another fact that we all know perfectly well from our own experience is that the very same thing holds true of conduct. It is not enough just to have memorized five moral lectures or sermons which admonish us to curb our greediness, malice or indolence. This memorization will not make us self-controlled, fair-minded or hard-working. What will help to make us self-controlled, fair-minded or hard-working are good examples set by others, and then ourselves practising and failing, and practising again, and failing again, but not quite so soon and so on. In matters of morals, as in the skills and arts, we learn first by being shown by others, then by being trained by others, naturally with some worded homily, praise and rebuke, and lastly by being trained by ourselves. It was partly because we knew in our bones that standards of conduct are not things that can be imparted by lectures, but only inculcated by example, training and self-training, that we found the idea ridiculous that there ought to exist, if virtue is to be teachable, professors and encyclopedias of these standards of conduct, and lecture courses and examinations in them. If the supposed professors were to instruct by dictating, they would no more teach Tommy to keep his temper than his father would teach him, by mere dictating, how to bicycle or play the piano.

This point, that virtues, like skills, crafts and arts, can indeed be taught, not by lectures alone, but by example and critically supervised practice, is a big part of Aristotle's reply to Socrates. Roughly, he says to Socrates, 'You are wrong to be surprised that, if virtue is teachable, there should exist no specialist lecturers on standards of conduct. For standards of conduct, like skills, are not bits of theory or doctrines, and so are not learned by the memorization of things dictated. We learn them by exercise. The things that we have to do, when we have learned to do them, we learn to do by doing them.'

Keeping one's temper is, in this respect, like playing the piano. The wisest theorists in the world can lecture as eloquently as you please; but the clearest memories of their doctrines will not, by themselves, enable Tommy to play the piano. A few weeks of five-finger exercises on the piano itself are needed to start with. *Mutatis mutandis* ditto for control of temper. Learning doctrines by heart is one sort of learning; learning to do things is another sort of learning, and one which is generally not much assisted by learning doctrines by heart. Learning to keep one's temper and learning to play the piano belong to the second sort of learning. Socrates' puzzle partly stemmed from his assuming that learning to keep one's temper belonged to the learning of the first sort, instead of to the second sort.

Well then, does Aristotle's answer to Socrates, married with Protagoras' answer to Socrates, solve the whole problem? I do not think that it does, though it is an important contribution to the general theory of learning and teaching. It emancipates us from the supposition that all lessons are dictations, and that all learning is learning by heart. Repertoires are acquired and abilities are acquired. But abilities are not repertoires. However, Socrates could still and should reply to a modern Aristotle in this way: 'I quite agree now that not all teaching is lecturing, and that teaching people to play the paino, bicycle, calculate, pronounce, design, translate, etc., largely consists in putting them through critically supervised training exercises in the operations themselves. Indeed, I grant that in real life even the most academic of teachers, namely tutors, demonstrators and professors in universities, spend much more of their time in teaching by training than they do in teaching by dictating—though the news of this fact does not seem to have reached the ears of the people who write to your *Times* newspaper about reforming your universities. But still my main puzzle remains. There do exist rowing coaches, swimming instructors, golf professionals, laboratory demonstrators and college tutors, who all teach people to do things chiefly by exercising them in the doing of these things. But again I ask: Where are the corresponding tutors, demonstrators and coaches in considerateness, toleration, pluck and candour? If flying, Latin prose composition, algebra and embroidery can be taught by critically supervised exercises, why cannot moral standards be inculcated in the same way? Or if they can, why do professional trainers in them not exist? Is it that moral standards are so elementary and so easy to pick up that they are left to be taught by any casual amateurs to children in their earliest years? Are they like skipping, snap, conversational English and hide-and-seek, which do not need, and unlike swimming, Latin prose and embroidery, which do need, expert instructors?'

We feel, I think, that Socrates' puzzle has still got something wrong with it. We do not really think that there should or could exist virtue professionals in the way in which there should and do exist golf professionals, swimming instructors and Latin prose tutors. But why not, if we continue to think that standards of conduct do have to be learned? What is the residual difference that we vaguely feel to exist between the acquisition of competence in snap, conversational English or algebra and the acquisition of conscientiousness?

Part of the difference is this, as both Plato and Aristotle clearly realized. A person may, by training, together with some native aptitude and zeal, become relatively good at something, e.g. relatively proficient in swimming, surveying or dissecting. But a proficiency can always be improperly as well as properly employed. The marksman who is able to hit the bull's-eye whenever or nearly whenever he wants is also able mutinously to miss it whenever he wants. The clever surgeon, who can repair internal lesions, has the skill necessary to cause fatal lesions, if the heirs to the patient's fortune can bribe him to do so. A bad, though well-meaning, engineer may by chance design a bridge that carries heavy traffic but breaks down in a gale. It takes a good, though unscrupulous, engineer deliberately to design a bridge that will carry heavy traffic and yet collapse in a gale. A man with the linguistic capacity to report clearly and coherently a complete budget of true tidings has the linguistic capacity to come out with a clear and coherent pack of lies if he wants to. What we object to about the surgeon who dexterously brings about the death of his patient is not any flaw in his surgical techniques, but the ends to which he exercises them. Technically we may admire an engineer who carefully designs an un-gale-proof bridge; morally we disapprove of him for wanting to sell his engineering skill to the ferry company whose trade will be ruined by a bridge. His techniques are good, but his motives are bad.

Virtues are not skills or proficiencies, since for any skill or proficiency it is always possible that a particular exercise of it be both technically first rate and unscrupulous. Beautiful specimens of carefully formed handwriting are to be found in the handiworks of forgers. Training a person in penmanship may make him successful as a forger, if he wants to make money dishonestly; it has no tendency by itself to make him not want to do this.

This brings out part of the underlying reason for our discomfort with the idea that there ought to exist experts to train us in standards of conduct, namely that having such standards of conduct is not being an expert in anything; so learning such standards of conduct is not acquiring expertness in anything. When, in writing a testimonial on behalf of a pupil, I have listed all the things that he is now highly or decently proficient at, I have still not said whether they would make it safe or dangerous to employ him, that is, whether he is scrupulous or unscrupulous, loyal or disloyal, straight or crooked. Being straight is not being an expert or even an amateur at anything; we cannot therefore teach people to be straight by just the kinds of teaching by which we train people to be proficient or competent at things. I do, of course, also say in my testimonial that 'he is a thoroughly loyal, modest and scrupulous man', but I do not say this while I am listing his abilities—or, of course, while I am listing his incompetences either.

We can now see the gap in Protagoras' reply to Socrates. You remember that he likened the child's learning of standards of conduct to the child's learning of his mother tongue. Neither derives from merely receiving informative lectures, and *a fortiori,* not from lectures by non-existent pundits of morality or of colloquial Greek. But now a big difference breaks out. Learning one's mother tongue is acquiring competences and skills. Of two children of the same age but with different aptitudes and different upbringings, one may be a lot more fluent, eloquent and coherent than the other. He is better than the other in colloquial English. But he may still be much worse than the other in his application of this proficiency. He may be a glib little liar, an eloquent little blabber of secrets, a coherent little carrier of malicious gossip. He has learned well how to tell things, but he has not learned what sorts of things he should and should not tell.

So a child can indeed pick up the main techniques of conversational English from the conversation of any English speakers with whom he associates; but this does not suffice for him to acquire standards of conduct, including standards of conversational conduct. Similarly, any competent player of snap can teach an ordinary child how to play snap; but it takes something quite different from competence at snap for a person to teach the child to prefer losing the game to winning it by cheating. So Protagoras is wrong if his analogy is meant to show that teaching a child not to want to cheat is all of a piece with teaching him how to play snap; or that teaching a child not to want to say spiteful things is all of a piece with teaching him how to say effectively whatever things, including spiteful things, he wants to say.

What, then, was the sort of teaching or training as a result of which young Jones, say, did grow up to be fairly or very plucky, considerate and trustworthy, if it was not just from sets of lecture notes, and also not just from the critically supervised practical exercises by which his various proficiencies were inculcated, tested and kept unrusty? As we have seen, these last may have made Jones a clever surgeon or engineer, but it was not these that made him not want his patient to die under his knife, or made him not want his bridge to collapse in a gale. So how was Jones taught or trained to want some sorts of things and not to want others; to aim at some sorts of goals and to shun others; to try to advance some sorts of causes and to despise others? If Jones is a conscientious surgeon, then his conscientiousness is no part of his dexterity, and vice versa; and the training that made him dexterous is not what made him care more for the welfare of his patient than for any other competing consideration that might be suggested to him. So how did he learn to care more for his patient's recovery than for any rival bonus? Unless we surrender and say that Jones was just born to be both asthmatic and conscientious, we seem now to be postulating a kind of learning by which he acquired not information and not proficiencies, but the caring for some sorts of things more than for others; a kind of schooling as a result of which, to put it in metaphor, Jones's heart came to be set on some things and against other things. He is now revolted by the suggestion that he or any other surgeon should for profit fudge an operation. But how can a person have been so instructed or trained that he now *feels* things that he would not otherwise have felt? Can there be lessons in feelings? Examinations in sentiments? Courses in attitudes? Drills or tests in motives?

We have now got Socrates' central question 'How, if at all, can virtue be taught?' separated off from questions about acquiring information and acquiring proficiencies; and the odd thing about this central question now is that in one way we all know the answer to it perfectly well. We remember how our parents reprimanded certain sorts of conduct in quite a different tone of voice from that in which they criticized or lamented our forgetfulness or our blunders. Some of the examples set us by our parents or by the heroes of our stories were underlined for us with a seriousness which was missing from the examples they set us that were merely technically enviable. The responsibilities given to us at school, if we had proved reliable, were of a different order from the tasks set merely to test us. We remember the growing, though inarticulate, distaste with which we witnessed the meanness of one schoolmate and the boastfulness of another. Above all, perhaps, we remember the total difference in gravity between the occasions when we were seriously punished and the occasions when we merely paid the routine penalties for infringing school—or family—regulations.

In these and countless affiliated ways we were, in a familiar sense of 'taught', taught to treat, and sincerely to treat, certain sorts of things as of overwhelming importance, and so, in the end, taught to care much more whether we had cheated or not than whether we had won the game or lost it; to care much more whether we were hurting the old dame's feelings than whether we were being badly bored at her little tea-party and so on. These are just familiar platitudes from the nursery and school.

But just to recite such platitudes as our answer to Socrates' central question, in its final shape, feels unsatisfactory because it seems to presuppose a general theoretical assumption which, in its nakedness, looks very suspect. The assumption is this: that we can properly be described as having learned or been taught standards of conduct when, under the influence of other people's examples, expressions, utterances, admonitions and disciplines, we too have come to care deeply about the things that they care deeply about. Facing this theoretical assumption, we are strongly tempted to object: 'Oh, but caring about something cannot properly be classed as something we have learned or been taught. Caring is just a special feeling that we have been conditioned to register on certain occasions. It is almost of a piece with the seasickness felt by a bad traveller the moment he steps on to the still stationary ship. We should never say that his previous unhappy voyages had *taught* him to feel *mal de mer* on embarking on a stationary ship. Your now feeling shocked by what used to shock your father or headmaster is just another effect of suggestion.'

What are the sources of our reluctance to accept in theory an idea which we accept unhesitatingly in daily life, namely the idea that people can be properly said to learn to want things, learn to admire things, learn to care about things, learn to treat things seriously in word, deed and tone of voice, learn to be revolted by things, learn to respect, approve and back things, learn to scorn and oppose things and so on? Whence, too, our reluctance to allow that non-moral preferences and partialities, like tastes and hobbies, may be not merely acquired, but acquired by cultivation? That today's Bridge enthusiast learned not only to play, but also to enjoy playing the game? One source is this. In our abstract theorizing about human nature we are still in the archaic habit of treating ourselves and all other human beings as animated department stores, in which the intellect is one department, the will is another department and the feelings a third department. Our poor bodies, of course, are not departments but basement kitchens, sewers or at best shop windows. So we take it for granted that as the intellect is notoriously the one deparment into which lessons go, our wills and feelings are not themselves teachable. They cannot know anything; they cannot be more or less cultured or cultivated. Alarms and hostilities cannot themselves be silly or sensible. Somehow or other our intellects can indeed harness, drive, steer, flog, coax, goad and curb our wills and our feelings; but in themselves these two brainless faculties are in the civilized man just what they are in the savage; there is no schooling of them. They are just brutes, like the docile horse and the undocile horse in Plato's *Phaedrus*.

This department store yarn is sheer fairy-story. It answers to almost nothing in the actual composition of human nature. Try for a minute to think of some friend of yours in terms of it; try to describe his actions and reactions during some recent crisis in terms borrowed from it, and you will see that a description which fits the department store yarn cannot be made to apply to your friend, and a description which fits your friend's behaviour cannot be wrenched into tallying with the department store yarn. No novelist, dramatist or biographer nowadays dreams of depicting his heroine or even his villain in department store terms.

Consider one example of a perfectly familiar feeling, namely amusement at a joke. Is it true, first of all, that you can separate off in it, first, an intellectual operation of considering the point of the joke and, secondly, an induced throb, spasm or pang of feeling tickled by it? Can you imagine having the tickle without having seen the point; or can you imagine seeing the point of the joke and feeling, just for once, not amusement, but vertigo, nausea, awe, anxiety or compunction instead? No, you want to say, and are quite right to want to say that the seeing the joke, the appreciating its wit or absurdity, the feeling tickled by it, and, unless in church, the laughing or smiling at it, are not separable operations or spasms, but are all features of the same thing, namely appreciating the joke.

Next, is it true that a sense of humour is something incapable of being schooled? Surely just the reverse. The infant in the cradle does not appreciate jokes; the toddler appreciates simple practical jokes, but not yet verbal witticisms; the schoolboy already appreciates puns, but not yet Oscar Wilde's epigrams; and so on. The sense of humour is an educable thing; certain sorts of jokes are above the heads of people of certain ages and of certain cultural levels. What greatly amuses a young man may show the quality and the academic level of his wits just as well as his essays for his tutor, or his comments on the news of the day. It may also show the quality of his character just as well as his manners and his remarks. But if a sense of humour is, in some measure, an educable thing, so that what tickles a savage or a schoolboy is more primitive than what tickled a Voltaire, then the partitions are down between the supposed departments in our department store; and then no architectural barrier survives to prevent us from saying that the reason why a suggestion that does not revolt a witch-doctor does revolt a surgeon is that the surgeon has what the witch-doctor lacks—the feelings of a civilized man. Only we must now cease to restrict the notion of *civilizing* to those of *dictating* and *coaching*. We must also cease to box up the notions of, for example, *caring* and *objecting* inside those of *sensations* and *sentiments*. For even the hospitable label 'feeling', which covers a huge variety of disparate things, overlaps only parts of such notions as *caring*, *objecting*, *taking seriously*, *being shocked at, championing,* etc. Seasickness and vertigo involve feelings of a purely physical sort, i.e. sensations. We feel these in our stomachs or heads. There is no question of our learning to feel sensations of these sorts. But amusement, indignation and compunction are not sensations. Only thinking beings can feel amused, indignant or penitent, and to feel these things they have to be, for example, appreciating the point of a joke, considering the injustice of an allegation or recollecting what they themselves had so inconsiderately done. To be educated up to the point at which we can recognize the injustice of an allegation is to be educated up to the point at which we can be and therefore, feel indignant about it, and vice versa.

But nor is indignation reducible merely to the having of a retrospective thought together with the synchronous registering of an induced sentiment. The indignant man is ready or eager to act indignantly, to sign his name to letters of protest, to voice indignantly his indignation in conversation with other people and perhaps to subscribe money for the vindication of the injured person. If Jones is indignant, then there beats indignantly the unpartitioned heart of a thinker of thoughts, an active performer of concrete deeds, a participant in conversations, an experiencer of sentiments, and perhaps a writer, years afterwards, of memoirs. His indignation lives in the ways in which he thinks, acts, speaks, swears, glowers, reminisces and even, perhaps, goes off his food and loses sleep. To describe him as caring deeply about the matter is not to report merely the occurrence in him of some induced, introspectable perturbation; it is to report how he takes the matter, where his 'taking the matter' embraces his thoughts, actions, words and facial expressions; it embraces the things he puts on one side as momentarily unimportant, the resolutions that he makes to do everything possible to get justice done, his gradings of his colleagues as staunch or as spineless and so on. If we like still to treat him as a department store, his indignation is not proprietary to just one of his three departments. It works from his head office down through all his departments, even to his sewers.

So to have acquired a virtue, for example, to have learned to be fairly honourable or self-controlled or industrious or considerate, is not a matter of having become well informed about anything; and it is not a matter of having come to know how to do anything. Indeed, conscientiousness does not very comfortably wear the label of 'knowledge' at all, since it is to *be* honourable, and not only or primarily to be knowledgeable about or efficient at anything; or it is to *be* self-controlled and not just to be clear-sighted about anything; or it is to *be* considerate and not merely to recollect from time to time that other people sometimes need help. Where Socrates was at fault was, I think, that he assumed that if virtue can be learned, then here, as elsewhere, the learning terminates in knowing. But here the learning terminates in being so-and-so, and only derivatively from this in knowing so-and-so—in an improvement of one's heart, and only derivatively from this in an improvement of one's head as well. A person learns to be honest or learns not to be a cheat; only derivatively do we need to bother to say he then knows that honesty is a good thing, or realizes that cheating is wrong. Certainly he does know or realize these things. But we think of him as a knower of such things chiefly when we think of him in his capacities as a relatively detached counsellor of others; as a preacher; as a penitent in the confessional; as a passer of judicious verdicts; as a legislator; that is, not in his capacity as a conscientious doer of things, but in his capacity as a source of precepts, confessions and tributes about the doings of things, i.e. as a spokesman or acknowledger of standards. He could not, indeed, be a spokesman for standards if he did not have them. But his having these standards is primarily his being fair-minded, considerate, self-controlled, etc., and only secondarily is it his being a reliable authority on these standards or an honest confessor to them.

When I have learned some dictatable things, like the dates of the kings of England, I may in time cease to know these dates, for I may forget them. When I have learned how to do something, e.g. compose correct Latin verses, I may, after a time, gradually lose this capacity, since I may, through lack of practice, get rustier and rustier. But a person who had once learned, say, to be considerate, though he may after a time become inconsiderate, will not naturally be described as having forgotten anything or got rusty at anything. It is not just knowledge that he has lost, whether knowledge *that*, or knowledge *how*. It is considerateness that he has lost; he has ceased to be considerate and not just ceased to know, say, some principles about considerateness. His heart has hardened, so it is not reminders or refresher courses that he needs.

Consider this parallel. To become good at bridge or cricket is not the same thing as to become keen on the game, though there is a natural connexion between the two. Losing one's taste for the game and becoming rusty at it are also different things, though either might be the result of the other. The acquisition of skills and keennesses can both be called 'learning'; but the losing of skills and keennesses cannot both be called 'forgetting'.

So if Socrates now asks us, 'Well then, who are the teachers from whom, on your view, young people learn to be straightforward, unspiteful, industrious and so on?', our answer will be partly a Protagorean one. We shall say something like this: 'To ask your question is to ask whom young people are apt to try to live up to; and the familiar answer is that many of them try, off and on, to live up to (a) some of the elders who bring them up, namely parents, uncles and aunts, elder brothers, parsons, schoolmasters, etc.; (b) some of the folk that they happen to live among or meet; and (c) some of the people that they hear about and read about. But most of these folk, being

quite unaware that anyone is trying to live up to them, could not, even with a stretch, be classed as 'teachers'. Nor, in actual cricket matches, was Bradman out to teach the cricket strokes that his young devotees were none the less sedulously learning from him. He was out to win his matches. He was unintentionally being their model; but he was not being their mentor.

Indeed, in matters of morality as distinct from techniques, good examples had better not be set with an edifying purpose. For such a would-be improving exhibition of, say, indignation would be an insincere exhibition; the vehemence of the denunciation would be a parent's, a pedagogue's or a pastor's histrionics. The example authentically set would be that of edifyingly shamming indignation. So it would be less hazardous to reword Socrates' original question and ask not 'Can virtue be taught?' but 'Can virtue be learned?' and to think less about the tests and techniques of instruction and more about the tasks and puzzlements of growing up.

Socrates might complain at this stage that the so-called learning that is now being postulated is only part of the instinctive imitativeness of the young human being, of the ape and of the parrot. The child's acquisition of fairmindedness, say, is all of a piece with his acquisition of a Yorkshire accent. Both are mere examples of being conditioned. Now it is certainly true that, without conditioning, the child will acquire neither conversational English, nor manners, nor morals nor a Yorkshire accent. Initially instinctive aping is a *sine qua non* of learning. But aping does not suffice to get the child to the higher stage of making and following new remarks in English, of behaving politely in a new situation, or of making allowances in a competitive game for a handicapped newcomer. He now has to *think* like his elder brother or the hero of his adventure story, i.e. to think for himself. He has now to emulate their non-echoings. No longer does he merely respond instinctively. He can now be silly, sensible and even clever; rude, decorous and even courteous; unsporting or even sporting. Now some of his acts, words, expressions and perturbations are misclassified as 'responses'. They need a personal pronoun in their description.

There remains a question which this paper does not try to answer. What of the young disciples of Fagin? They live up to his standards; they dread his frowns; they learn from him to scorn mercy, unworldliness and good citizenship; they are, perhaps, as scandalized, revolted or indignant as he would be at exhibitions of scandalization, revulsion or indignation. Are what they have learned from Fagin virtues? We can think of some of our own acquaintances who speak of toughness as a virtue, and others who speak of it as a vice, and vice-versa of meekness. How should we decide who is right? Why need the list of good and bad qualities of character on which I draw when writing testimonials tally wholly or even partly with the lists drawn on by Confucius, Nietzsche or Machiavelli? Cannot vices be learned?

14. Moral Education in the Secular School

Paul H. Hirst

* * *

THE CASE FOR EXPLICIT MORAL EDUCATION

Moral education cannot, of course, simply be left to the general influence of the school, even if that includes a properly constituted way in which the individual's interests are fully safeguarded. Adequate education in this area, as in any other which involves a great deal of understanding and intellectual mastery, necessitates much explicit learning. Somewhere, somehow, every school curriculum should provide opportunity for pupils to acquire the very considerable amount of knowledge that is necessary for morally responsible living in our complex democratic society, and the intellectual skills and dispositions the making of moral judgments demands. Where sheer knowledge of facts about the world and society is concerned, it is remarkable how little care we take to make sure that school-leavers are well-informed for the responsibilities they must take on. Whether one is talking about such personal matters as, say, sex and drugs, or civic matters about the law and the working of political institutions, we have still not started to get to grips with producing curricula that are responsibly planned. And that in a situation where society progressively gives young people more and more personal freedom and more and more social and political power. Freedom and power in the hands of people who are simply ignorant of what they are involved in are hardly likely to produce the good life for anyone. In both the natural sciences and social studies we need to map out for all pupils curricula that no longer skirt this issue. We are still addicted to the idea that pupils

Reprinted with permission of the publisher, Hodder & Stoughton, from *Moral Education in a Secular Society* (London: University of London Press LTD, 1974), pp. 108–116.

learn what is important by in fact learning something else. The understanding relevant to teenagers' problems of sex and drugs that actually emerges from biology lessons is often very limited and dangerous in its partiality. The understanding as to how public institutions in our society actually work that emerges from the study of British history, the nearest many get to the matter, is usually minimal. Do we not care that so many schools do nothing to enable pupils to understand the contemporary significance of, say, trade unions, local government, or public finance?

Knowledge of this factual kind is, however, far from adequate for the making of moral judgments in both the personal and public areas of life. An understanding of the attitudes, feelings, values and motives of people is equally important. In school, some of this comes from the direct experience of living with others in the school, but that is necessarily so limited that much more is required. In the study of literature, history and religion there is a great opportunity for pupils to understand 'how people tick', to enter imaginatively into situations of great diversity. It needs to be more widely recognised, however, that this understanding of people is indeed what education is after in these subjects, at least in part. Only too often very little of this kind is sought and even less gained. Maybe, if it is understanding people we want, other media have much to offer: film, drama, TV. And yet more, perhaps we must again go directly for what we want by extending pupils' personal experience, educating them within situations that communicate face to face other attitudes, values and so on. Psychologists have had much to say here about the need for pupils to come to understand themselves, their own feelings, attitudes and values.[1] Achieving a sense of personal identity, as someone who counts in the eyes of others, has a significant role to play in the community, who has enough confidence to express a coherent personal point of view, all this matters greatly. No doubt the life of the school outside curriculum areas is the context in which much of this sense of oneself develops, if only because of the diverse opportunities for experiment and role-taking that are possible. Nevertheless, the curriculum itself can offer considerable scope as well. New activities for achieving personal understanding need to be built into school work and, if many of these are not likely to be highly theoretical in character, there is no doubt a place too for responsible teaching in elementary psychology and sociology.

These critical suggestions inevitably lead to the question as to whether moral education should enter curricula as a distinct subject or area of study. Though much of the understanding I have just been discussing can be catered for within other curriculum units, and to my mind should be so handled, I think there is nevertheless an overwhelming case for bringing together the relevant knowledge in periods concerned explicitly with moral education. Every curriculum area deals with matters related to some moral problems, but most cannot deal very readily with all the different matters relevant to many complex issues without deliberately devoting considerable time to them. If, however, one recognises that learning to make moral judgments involves learning to ask particular kinds of questions and to reason in particular ways, then the case for explicit attention to moral questions in time set aside specially for this work becomes very strong indeed. Not only is it then recognisably one person's responsibility to see that the necessary knowledge is in fact acquired by pupils, it is his responsibility also that they reach an appropriate mastery of the logic of moral discourse and a grasp of fundamental moral principles. The whole question of pupils grasping the significance of public and private morality in society, the relationship between morals and religion, law and convention, all these are matters adequate education must take on at an appropriate time. Periods allocated for this purpose seem to me inevitable. Not that any aspect of moral education should be left to such periods only; if a highly suitable opportunity presents itself for dealing with the matter on some other occasion it should be used. Flexibility is of the essence of good education, but within a recognised scheme of responsibilities. Nor am I suggesting that periods set aside for this work should be labelled 'M.E.' Labels are immaterial as long as they are not misleading, and periods set aside for moral education can readily occur within areas designated by the labels of topics dealt with or under such headings as 'modern society' or 'humanities'. I would, however, myself resist explicit work in moral education being swept up under the label of an existing conventional school subject that would be misleading.

In particular, it seems to me undesirable to deal with moral education under 'English'. That linguistic and literary study has much to contribute to moral education is, of course, true. That conventional syllabuses in this subject can readily include many other elements necessary to moral education I would reject. The pretensions of certain literary critics and teachers of English have done much to confuse, rather than elucidate, the nature of moral issues, and have encouraged many teachers quite unqualified for the job to undertake a dangerously inadequate approach to moral education through the use of English literature.[2] Nor do I think it appropriate that moral education be simply subsumed under 'religious education' or 'religious studies'. Such a procedure, whatever the form of teaching it covers, is of itself liable to suggest that morality is not autonomous, when its autonomy is one of the central elements that needs to be taught. If religious education is being conducted with the aims I have suggested are appropriate for the secular school, an understanding of the relationship between certain religious positions and certain moral beliefs and practices will be included. This involves recognising, however, that moral education is in a very different position from religious education as commitment to certain rational moral principles is thoroughly justifiable. The central concern of the school with moral education should therefore be dissociated from religious education, it being accepted that only by some is morality thought to have a religious relevance and that that relationship is in any case irrelevant to the justification of moral principles.

In insisting on the need for specific attention to moral education in the curriculum, I have tried not to deny the significance of all other areas of the curriculum for this work. Every element of the curriculum will involve pupils in learning forms of thought as well as a content which have their place in moral understanding. All teaching contexts too involve relationships and patterns of behaviour that are significant in learning moral conduct. It is the adequacy of any and all of these contexts for this particular job that I am calling in question. I am suggesting too that no teacher trained in any established academic discipline, and skilled in teaching it, is thereby knowledgeable about either the nature of moral problems and their solution or the best ways of carrying out moral education. What we need are teachers who have studied the nature of morality, who have the necessary contributory forms of understanding on an appropriate area of moral questions, and are trained to teach in this area. There are at present few such people available. That there should be teachers with this specialism is to many a sinister proposal, for must not all members of the profession be morally educated to a degree that would qualify them to do this job, and is one not in danger of producing moral authorities of a dangerous kind? That all teachers are expected to be educated sufficiently to use English in an accurate and appropriate way is not regarded as a reason for denying the necessity for specialist teachers in this area. Most teachers would be quite inadequate in the area, possessing none of the necessary analytical grasp of the subject matter or of its presentation. Much the same must be true where morality is concerned. Nor are moral specialists to be feared if they are indeed aware of the real nature of morality and of the importance of rational autonomy. It is the evangelist, or the 'specialist' who does not understand the nature of the subject matter, who is the menace. But such people are a menace in every area and that fact cannot be allowed, in moral education any more than elsewhere, to cloud the question as to what the proper conduct of the professional task necessarily demands: teachers adequately educated in what they teach and trained to educate others accordingly.

TEACHING MORALITY

If it is agreed that the school curriculum should specifically deal with morality, what precisely should be done in the time available? From an intellectual point of view, there must be a place for developing the many elements of necessary knowledge and understanding that I have referred to repeatedly. This can be done in a great variety of ways and I shall not comment on this further. In particular, of course, there must be attention to moral problems themselves. Which problems at which stage depends on many factors. It is possible to present for discussion written work, dramatic expression and so on, life situations which are appropriate to the age and interests of pupils. In this way relevant matters of fact and of principle can be harnessed to the making of judgment. New concepts can be developed and new forms of reasoning can be practised, much of this in learning to use and apply moral discourse. But if this is to be effective, Kohlberg's stages must be recognised and the situations provide at least intellectually for the role-taking he insists is crucial. There is, however, every reason to think that if dispositions to think morally are to be developed, the situations and problems used must be of genuine interest to pupils, involve their emotions, their imagination, and their desire to find solutions that would make practical sense for their own actions. It must be the real life of the pupils that is focused on, not the real life of adults.[3] It is also a question of dealing with these, not in detached terms such as 'what ought to be done about racial discrimination in this case?' but rather in terms of 'what ought we to do about this case of racial discrimination we know about?' And that immediately introduces a further point, that if dispositions to act rationally matter as much as dispositions to think rationally, the process of encouraging responsible moral judgments without related action is of itself inadequate and might well encourage an undesirable divorce of moral thought and action. It seems then that at the centre of explicit moral education there should be the study of, and involvement of the pupils in, particular moral activities that they are able to see as important. That means, on the one hand, involving them in the detail of the life of the school community and using that fully as an educational instrument both theoretically and practically. On the other hand, it means involving pupils significantly in moral issues in the society outside the school, again not only theoretically, but in action as well. If both these are done, they will provide opportunity too for the development of social skills without which the moral life can be only too easily vitiated. It is encouraging that there are now becoming available both materials for use in school and suggestions for relevant activities both inside and outside the school that will help just this kind of work.[4]

There is a tendency to think of moral education as simply a matter of the discussion of suitable material, and it must be apparent that that is to my mind, on its own, much too limited an approach. Nevertheless, discussion must be a vital part of the process provided it is seen as a teaching method. The airing by pupils of their ignorance and prejudice whilst a teacher assumes a form of moral neutrality is a travesty of education.[5] The open-ended nature of debate on controversial matters must indeed be recognised when such matters are discussed, but that that is true of all moral issues cannot, on my argument, be held by any rational person. If it is the job of the teacher to promote rational autonomy, discussion is useful as a method only in so far as it promotes that end. Let us by all means have much more discussion work of this kind, but let us have it geared to the development of reason and indeed to the development of rational action. In that case, the method will be characterised neither by an irresponsible acceptance of pupils' autonomy uncontrolled by reason, nor by an uncritical indoctrinatory imposition of the moral opinions of the teacher.[6] In moral education, as in any other area of education, what is asked of the teacher is a total commitment to the development of rational autonomy in both thought and action. Teaching that begins to suggest that any beliefs cannot be rationally called in question, or that seeks to develop dispositions against such questioning, is not acceptable. But to say that is perfectly consistent with also saying that in morals, as in mathematics, history, or any other area, there may be a body of beliefs and principles that can be taught with confidence as having substantial rational defence. It is the job of the professional teacher to know the limits of that defence and to be true to those limits. Where reason can offer no conclusions, education can offer none either. Where reason can offer conclusions, even if they are provisional as in so many matters, to pretend otherwise is unreasonable. What education demands is that in content, teaching methods, discipline and relationships within the school, teachers be governed by nothing but reason.

The central problem in education today seems to be a failure of nerve on just this very point. Not only moral education but education as a whole seems to be losing some of its bearings in moving from a primitive concept to an unqualified concern for reason. Of course, such failure of nerve is understandable when society is undergoing so many changes, when reason has made such mistakes in the past and has proved so powerless against the forces of unreason. No wonder we neither stick consistently to principles and practices for which the rational defence is abundant, nor follow consistently new lines of development where reason manifestly demands them. But this failure of nerve must be overcome if educational, and indeed social, disaster is to be avoided. After all, there is no basis other than reason for meaningful human development. Both personal and social salvation may to the Christian have their source in God, but I see no grounds for thinking that, even on that view, human reason can properly be set on one side. The secular society is supremely the product of reason, God-given reason if you will. Its problems come not from the development of reason, but from our refusal on so many fronts actually to live accordingly. There are today very few major social issues on which the most disruptive elements in our society have not got a highly defensible reasoned case. True, irrational elements may repeatedly take over the cause, but they take over the defence of the *status quo* as well. We do not live in a society that is morally defensible in its distribution of power, wealth, or, for that matter, true education. We deceive ourselves if we think otherwise. The increase of secularisation with its attendant growth of reason has made this plain to a point where it can no longer be hidden. What we shall have to do, if our society is not to become morally degenerate and return to control by force, is re-fashion it so that reason can in fact prevail. To this end, moral education in schools has something to contribute, if alone it is powerless against other social institutions. We need the next generation to be more moral than we are, and that means more committed to reason. I am not altogether without hope, provided we can all, Christians and non-Christians alike, stop seeking irrational solutions to our ills and produce education for rational autonomy. That alone is the form of moral education that can properly serve our secular society.

NOTES

1. WILSON, J. (1972). *Practical Methods of Moral Education.* London: Heinemann.
2. The background to much of the advocacy of moral education through literary study is to be found in the writings of F. R. Leavis. See particularly his *Education and the University* (Chatto and Windus, 1948). A critical study of this view of literature and its historical origins is to be found in: WILLIAMS, R. (1961). *Culture and Society.* Harmondsworth: Penguin Books.
3. See MCPHAIL, P., UNGOED-THOMAS, J. R., CHAPMAN, H. (1972). *Moral Education in the Secondary School.* Harlow: Longman Group for Schools Council.
4. See particularly:
   MCPHAIL, P. (1972). *In Other People's Shoes,* Harlow: Longman Group for Schools Council. Also materials of the Schools Council Project in Moral Education published under the title *Lifeline.*
   (1970). *Schools Council and Nuffield Humanities Project: An Introduction.* Heinemann Educational. Also handbooks and materials developed by the Project. Community Service Volunteers: materials produced under the general direction of Dickson, A.
5. See BAILEY, C. (1973). 'Teaching by Discussion and the Neutral Teachers'; and ELLIOTT, J. (1973). 'Neutrality, Rationality and the Role

of the Teacher', both in *Proceedings of the Philosophy of Education Society of Great Britain,* Volume VII, Number 1.

6. The concept of indoctrination has been widely discussed by philosophers. See: SNOOK. I. H. (ed.) 1972). *Concepts of Indoctrination.* London: Routledge and Kegan Paul.

15. Reflections on Moral Education

Peter F. Carbone, Jr.

One of the curious things about moral education is that while nearly everyone approves of it, we seem to have great difficulty in working it into our educational system. As Ralph Barton Perry observed some years ago:

> Schools and colleges, designed for educational purposes, leave it to the home, the church, the Boy or Girl Scouts, or other private and more or less impromptu organizations. But even these agencies hesitate to assume responsibility. The home passes it on to the school, and the school passes it back to the home.[1]

The literature on the subject, moreover, clearly tends toward the view that what little time and effort the school *does* invest in moral education is relatively unavailing. On that account, there is certainly no scarcity of articles pointing out the contemporary "breakdown" of moral standards and urging upon the schools the obligation to revitalize the nation's moral strength, the implication being, of course, that educators are not performing the task satisfactorily at present.

It seems to me that whether or not one subscribes to this view depends in large part on one's conception of moral education. It is doubtless true that we rarely allot a place in the curriculum for a formal course in the subject. Nor, as a rule, do we get very deeply into moral issues, even when we do attempt to provide at least a smattering of moral education on an informal basis. On the other hand, any experienced teacher can testify that the school takes some pains to reinforce those norms and values that are generally

"For the Record," in December, touched on the question of moral education. We are pleased to have Professor Carbone carry the discussion further. Not only does he speak here of the need to confront moral issues in the classroom; he talks of the specific problems entailed by the need to socialize young children into the moral institution, and the need to engage older children with critical ethical inquiry. An earlier version of this paper was presented at the annual meeting of the South Atlantic Philosophy of Education Society in October 1968.

Reprinted, by permission, from *Teachers College Record* 71 (May 1970): 598–606.

accepted in society at large. This is usually accomplished not by direct instruction in moral precepts, but rather by various indirect methods which stress example and illustration in a variety of contexts. The means employed are diverse and somewhat haphazard, perhaps, but the task can hardly be said to be ignored. It can, however, and frequently is said to be ineffective, but here again the assertion is somewhat ambiguous. If our criterion of effectiveness is the child's ability to recite the values and norms he is expected to abide by as the result of moral instruction, then I should say that the school, in conjunction with home and church, is fairly successful. For how many school children would deny that they should be God-fearing and patriotic; that they should tell the truth, be honest, and keep their promises; that they should love and respect their fellow men (communists, anarchists, and miscellaneous "leftists" excepted, of course); that they should value liberty, equality, and, above all, free enterprise?

APPROPRIATION AND INDOCTRINATION

The charge of ineffectiveness may refer, however, to actions, to what children do as opposed to what they say, in which case the criticism could be well-taken. For as Scheffler has so ably pointed out, it is one thing to appropriate a norm in the verbal sense and quite another to possess a tendency to act in accordance with it.[2] This being the case, it might seem at first glance that the solution lies in forging patterns of behavior consistent with the normative principles we wish to impart, using whatever behavior-influencing devices we may have at our disposal. Now the obvious objection to this strategy is that it smacks of indoctrination, and as Frankena notes after considering techniques along these lines, "We conceive ourselves as having put them behind us."[3] And so we have—but not completely, of course. Here again, those most familiar with what takes place in our classrooms would concede, I believe, that this sort of thing is hardly unknown in the American school.

Of course the indoctrination charge may refer to the content as well as to the process of moral education. This issue obviously emerges when we raise questions about which moral principles we should be expected to teach. As I indicated earlier, we do present long-standing norms and values to children as being worthy of adoption. This is part of what is meant by passing on the cultural heritage. But it will not do to construe the transmission of culture as the whole of moral education, since it is always appropriate—in fact it is incumbent when one is engaged in moral inquiry—to question the legitimacy of custom (and, indeed, of law or any other guide to conduct), and it is this feature more than any other, perhaps, that sets ethics or critical morality off from custom or conventional morality. In other words, an individual might understand perfectly well which norms are valued in his culture and yet reject some of them on the grounds that they are unacceptable from the moral point of view. As Benn and Peters have observed in this connection, "Morality arises when custom or law is subjected to critical examination."[4]

VALUE CONFLICTS

Thus, it is inappropriate, at least with older children, to teach morality the way we teach the multiplication tables, or the characteristics of chemical elements, or, for that matter, the behavior of crowds. Morality is not primarily an "information-dispensing" subject, in which content can be distributed in neat factual packages. "What distinguishes morality from the formal and natural sciences," says R. F. Atkinson, "is that in it different and opposed first principles are readily conceivable, and are in fact accepted by morally serious people."[5] It is important, I think, for students to grasp this fact. Similarly, it is important for them to realize that in a given situation dispute is entirely possible, even among those who subscribe to the *same* first principles. As Isaiah Berlin reminds us:

> In life as normally lived, the ideals of one society and culture clash with those of another, and at times come into conflict within the same society and, often enough, within the moral experience of a single individual; . . .[6]

This point seems to have been missed by those writers on moral education who exhort us to present prevailing norms and values to children as though we *were* teaching the multiplication tables, as though moral principles, like the rules of mathematics, never conflict with one another. Apart from the indoctrination issue, it is worth noting in this connection that even if we decided to heed this advice and ignore the problem of validation, we would still fall short of the mark from a practical standpoint. We would fall short because our students would be unprepared to cope with situations involving conflict between values. That such conflict is not only possible but rather commonplace needs to be clearly understood.

More than that, it is important, I believe, to emphasize that considerable disagreement exists with regard to the very nature of ethical propositions. Consider, for instance, the claim that a given act is right, wrong, or obligatory; or that a certain character trait or motive is morally good or bad; or that an experience or a material object is valuable (in a nonmoral sense). Can such claims be said to be true or false? Are they even meaningful? Are they empirically verifiable or logically demonstrable? Are they self-evident, that is, can their validity be seen intuitively? Or are such judgments merely matters of personal taste or opinion? Can they be described as purely subjective, emotional utterances? Or, finally, do they function neither as descriptive propositions nor as arbitrary assertions of personal preference, but rather as prescriptions which can be defended on rational grounds, though not verified to the degree possible with empirical or logical propositions? Turning to moral philosophers for guidance, we find, alas, that most if not all of these questions have been answered both affirmatively and negatively by competent thinkers.

FROM ACCEPTANCE TO CRITICISM

To grasp this characteristic open-endedness, to understand that there are no absolute, invulnerable guidelines to the "virtuous life," that no moral theory has preempted the field, is to begin to discern something about the structure of morality; and assuming that such discernment is helpful when one engages in moral discourse, I should think that it ought to rank high on our list of priorities. I am not suggesting, however, that we can dispense entirely with the inculcation of norms. The fact of the matter is that we cannot wait until the child has reached the point at which he is capable of abstract reasoning before we begin to introduce him to the norms and values that are part of his cultural legacy. The school could "officially" disclaim all responsibility for providing moral instruction in the lower grades, of course, but teachers would continue to impart norms in one way or another simply by virtue of their roles as authority figures in the lives of younger children. Thus it is unrealistic to argue that we can avoid confronting the inculcation bugbear merely by postponing moral education until the child is old enough to benefit from a more sophisticated treatment of the subject. The real issue is not whether, but how moral instruction should be provided in the elementary grades; and since the research of Piaget and his colleagues[7] indicates that youngsters at that age are incapable of grasping the rationale for moral principles, or even perceiving the appropriateness of demands for justifying reasons, it would appear that we are left with little choice at this stage but to present the rules as though they were part of the natural order of things. I am not overlooking, in this context, the heuristic educational value, particularly in the "factual" areas of the curriculum, of Bruner's interesting claim that "any subject can be taught effectively in some intellectually honest form to any child at any stage of development."[8] Given Piaget's findings, however, it is not at all obvious that this principle can be applied to the moral education of the very young without placing undue strain on the term "intellectually honest." The problem, then, is to avoid destroying the child's capacity for later critical evaluation of the norms he has been led to accept uncritically during his most formative years. Referring to this situation as "the paradox of moral education," Peters has described it as follows:

> Given that it is desirable to develop people who conduct themselves rationally, intelligently, and with a fair degree of spontaneity, the brute facts of child development reveal that at the most formative years of a child's development he is incapable of this form of life and impervious of the proper manner of passing it on.[9]

It is necessary, in short, to instill norms and habits of behavior before children are capable of thoughtful appraisal of what they are absorbing. Now it is very difficult for us to admit this necessity because the admission is so much at variance with our popular ideology. Indeed, a good deal of our educational rhetoric is utilized to deny this very assertion. The danger here, it seems to me, is that we can get so caught up in our own rhetoric that we fail to perceive the extent of, and the reasons for, our involvement in the practice of inculcation. Consequently, we tend to obscure the difficult problem of how norms may be implanted in children and yet not so firmly rooted that they will be permanently immovable under any contingency whatsoever. If, as Piaget and Peters seem to suggest, some inculcation is inevitable in the lower grades, then the question is, how much of it can we tolerate, and how can we avert its potential adverse effects? Hopefully, we may turn to the educational psychologist for assistance here, but such help is not likely to be forthcoming unless we ask the right questions. And in order to do that we must first face up to the problem.

THE SOCIALIZATION PHASE

What I am suggesting, then, is that it might prove fruitful to conceive of moral education as including two fairly well-defined levels or phases. At the first level our chief concern should be to contribute to the socialization of the child by inducing him to accept (in the active sense) the values, attitudes, and standards of behavior that prevail in his social environment. (There will be some conflict here, of course, but even in a society as pluralistic as ours, there are basic values that transcend group differences.) As I have already indicated, a certain amount of imposition is unavoidable at this level, there being no other way to initiate the young into their culture at the time such initiation must begin. By "imposition" or "inculcation" I do not mean what Sidney Hook calls "irrational" means of persuasion such as the systematic use of spurious arguments, for example, or the suppression of pertinent facts in order to support a debatable point of view.[10] This sort of approach is to be avoided at all levels. On the other hand, what Hook refers to as "conditioning" or "nonrational methods of inducing belief," presenting norms straightout, that is, without benefit of elaborate supporting statements or possible counter arguments, seems to me to be an acceptable method of instructing younger children who are not yet proficient in dealing with abstractions. Even at this early stage, however, we need, as Hook cautions, to be alert to opportunities for cultivating the child's critical abilities, and we should present to him on a nonrational basis only those norms that we are convinced will stand the test of reflective evaluation later on.

TOWARDS REFLECTIVENESS

In the second phase of moral education, our objective is, of course, to advance the child beyond the level of relatively passive acceptance of norms, merely because they are prevalent in his surroundings, to a point at which he is capable of critical, independent judgment in these matters. In a word, we are interested at this level in developing reflective moral agents, people capable of furnishing a reasoned justification for the principles that guide their behavior. For as Frankena comments,

> Morality fosters or even calls for the use of reason and for a kind of autonomy on the part of the individual, asking him, when mature and normal, to make his own decisions, though possibly with someone's advice, and even stimulating him to think out the principles or goals in the light of which he is to make his decisions.[11]

A good deal of the literature on moral education centers on the problem of how best to carry out what I prefer to think of as the preliminary part of the task. There is much debate about what means are most effective, whether, for instance, time should be set aside for direct instruction in moral precepts, or whether the so-called indirect methods—teacher example, illustrations drawn from the study of literature and the social sciences, inspirational school assemblies, object lessons arising out of classroom or extracurricular activities, etc.—will yield better results.

These are questions worthy of serious consideration, certainly, but a more important issue in my view, as I have already intimated, and one that does not usually receive the attention it deserves, is the problem of how to facilitate the child's transition from the first to the second phase of moral education. For surely we cannot rest content with simply furnishing instruction in whatever moral principles happen to prevail at present. Surely a second phase is needed if we take seriously the goal of producing autonomous moral agents. To continue on indefinitely with the techniques appropriate at the first level, I should say, is to fail to advance from moral "training" to "teaching," both of which have their place in an overall program of moral education. An adequate analysis of this distinction would take us far afield, but roughly, "teaching" is more restrictive than "training" in terms of acceptable methodology, and it demands more of a cognitive emphasis on the part of both teacher and learner. "To teach, in the standard sense," Scheffler remarks,

> is at some points at least to submit oneself to the understanding and independent judgment of the pupil, to his demand for reasons, to his sense of what constitutes an adequate explanation . . . Teaching, in this way, requires us to reveal our reasons to the student and, by so doing, to submit them to his evaluation and criticism.[12]

"Training" on the other hand, connotes processes of drill, rote-learning, habit-formation, and the like. It is more permissive, less scrupulous about the means used to bring about a change in behavior or in attitude. It fails, in sum, to engage the child's rational capacities to the extent that "teaching" does, and is therefore unequal to the assignment once moral education has advanced beyond the introductory stage.

TEACHING PRINCIPLES

Much more could, and no doubt should, be said by way of clarification here, but perhaps the point regarding the difference in emphasis between the first and second levels is evident at least in outline form. I have already commented on the need for further research to inform our efforts with respect to the preliminary phase. Assuming that this additional information will be provided, and that we *can* guide the child through his early moral training without placing too great a strain on his incipient critical capacities, we can sketch in some of the characteristics of moral education at the second level. Most of these characteristics, e.g., the emphasis on reasons and justification, the awareness that moral principles frequently conflict with one another, the realization that there is considerable disagreement even among moral philosophers concerning the meaning and cognitive status of moral propositions, have already been mentioned. In addition, we need to convey something about the nature of moral discourse, I should think, and perhaps some understanding of how it differs from the "language" of other disciplines. This, I suggest, is partly what we are groping for when we talk about providing children with the intellectual tools that are a prerequisite for clear thinking. Further, we need to confront our students with moral issues that force them to re-examine and re-evaluate their own moral principles. "What we do, if we are sensible," Hare writes,

> is to give him [the learner] a solid base of principles, but at the same time ample opportunity of making the decisions upon which these principles are based, and by which they are modified, improved, adapted to changed circumstances, or even abandoned if they become entirely unsuited to a new environment.[13]

Admittedly, it is somewhat unsettling to subject one's basic moral beliefs to the kind of challenge implied here; but, as Peirce and Dewey taught, the irritation of doubt frequently serves as a prod to genuine inquiry. It may be argued, however, that youngsters of high school age are not sufficiently experienced or psychologically stable enough to engage in this sort of thing, that such considerations should be taken up only on the college level.[14] Personally, I feel that this view grossly underestimates the maturity of contemporary 16-, 17-, and 18-year olds, who have practically been weaned on moral controversy as a result of their constant exposure to the mass media. And though it may be true that adolescence is not the most psychologically tranquil period in one's life, it is also true that it is the time when one is most likely to seriously question and demand justification for the moral rules and standards one is expected to honor. Under these circumstances, we do the adolescent no favor in attempting to shield him from difficult moral issues at a time when he is searching for a personal philosophy of life. What he needs at this point is guidance on a journey that he is very likely determined to undertake, whether we approve or not.

Most of the procedures suggested above are rather familiar, to be sure, yet with possible rare exceptions, one does not find them being implemented in our schools. Their absence is partly attributable, in my opinion, to the misconception that once the notion of moral absolutes is discarded, *any* attempt to provide moral education becomes an exercise in indoctrination. And while we might be willing to concede that a limited amount of indoctrination in Hook's "nonrational" sense may be unavoidable in the lower grades, most of us rightfully have serious misgivings about extending its application to older students. Thus we simply neglect to provide a coherent program for this age group. In my view this is a classic example of throwing out the baby with the bath water. The perceived danger can easily be averted by recognizing that at the second level our primary emphasis must shift from the transmittal of moral propositions to their application, justification, meaning, and genre. I am inclined to believe that such a shift in emphasis is mandatory if moral education is to reflect the essential character of moral philosophy, and that some such reflection is necessary to ensure the integrity of moral education.

In concluding, I should acknowledge the many practical problems that I have neglected to consider in this brief essay, problems relating, for example, to implications for teacher education, to the possible introduction of new courses, and to the relationship of moral education to the "factual areas" of the curriculum (for obviously factual information is a necessary condition for the intelligent application of moral principles). I realize, too, that some of the concepts and terms employed in this discussion would benefit from further explication and analysis. But each of these tasks would require extended treatment, and my purpose here was merely to suggest a concept of moral education which might contain enough initial plausibility to be taken up for further discussion. In attempting to do so, I have drawn freely from the writings of a number of philosophers who have made significant contributions to our understanding of the ways in which moral philosophy is relevant to the problems of moral education. If drawing together some of the more promising and provocative features of their work contributes anything worthwhile toward the development of an adequate conception of moral education, I am confident that the practical problems can be worked out by specialists in the areas of learning theory and curriculum development.

NOTES

1. Ralph Barton Perry, *Realms of Value* (Cambridge: Harvard University Press, 1954), p. 428.
2. Israel Scheffler, *The Language of Education* (Springfield, Ill.: Charles C. Thomas, 1960), pp. 78–86.
3. William K. Frankena, "Toward a Philosophy of Moral Education," *Harvard Educational Review* 28 (Fall 1958): 302.
4. S. I. Benn and R. S. Peters, *Social Principles and the Democratic State* (London: George Allen & Unwin Ltd., 1959), p. 31.

5. R. F. Atkinson, "Instruction and Indoctrination," in *Philosophical Analysis and Education*, ed. Reginald D. Archambault (New York: The Humanities Press, 1965), p. 176.
6. Isaiah Berlin, "Equality," *Proceedings of the Aristotelian Society* 56, 1955–56 (London: Harrison & Sons, 1956): 319.
7. Jean Piaget et al., *The Moral Judgment of the Child* (New York: The Free Press, 1965).
8. Jerome S. Bruner, *The Process of Education* (Cambridge: Harvard University Press, 1963), p. 33.
9. R. S. Peters, "Reason and Habit: The Paradox of Moral Education," in *Moral Education in a Changing Society*, ed. W. R. Niblett (London: Faber and Faber, Ltd., 1963), p. 54.
10. Sidney Hook, *Education for Modern Man: A New Perspective* (New York: Alfred A. Knopf, 1963), p. 170.
11. William K. Frankena, *Ethics* (Englewood Cliffs, N. J.: Prentice-Hall, Inc., 1963), p. 7.
12. Scheffler, *The Language of Education*, p. 57.
13. R. M. Hare, *The Language of Morals* (Oxford: Clarendon Press, 1952), p. 76.
14. This view is expressed in George Herbert Palmer and Alice Freeman Palmer, *The Teacher* (Boston: Houghton Mifflin Co., 1908), pp. 38–46.

Part Six: Moral Education: Theory Into Practice

Introduction

In this final section of the book, we shall continue our discussion of moral education with an increased emphasis on translating theory into practice. To that end, we shall examine the two most influential developments in recent moral education: the values clarification approach led by Sidney Simon and his colleagues, and Lawrence Kohlberg's cognitive-developmental movement.

We begin Part Six with excerpts from *Values and Teaching*, which is generally regarded as the definitive account of the values clarification philosophy. In these excerpts Raths, Harmin, and Simon argue that value judgments are neither true nor false. As products of personal experience, values are matters of individual choice. Since values cannot be held to be right or wrong, moreover, the emphasis in moral education should be on helping students to clarify their own values, rather than on persuading them to adopt ours. This approach, according to the authors, fosters interest, freedom of choice, and clear, critical thinking in students. Since it is based on respect for, and trust in, children, it also results in a classroom environment conducive to learning.

John Stewart's essay is highly critical of the values clarification approach to moral education. He considers values clarification to be both superficial and confused—superficial because it ignores the structure or "underlying cognitive logic" of value judgments, and confused because it concedes the existence of certain moral absolutes, a concession that undermines its theoretical relativism. In addition, Stewart thinks that values clarification theorists are mistaken in assuming that their process is free of content. In point of fact, Stewart argues, a number of values are clearly evident in various accounts of their own methodology. Thus from Doctor Stewart's point of view, values clarification turns out, somewhat paradoxically, to be a rather judgmental approach to moral education.

Lawrence Kohlberg and his followers have provided several accounts of their "cognitive-developmental approach to moral education" over the past twenty-five years or so. The essay in this section was chosen over possible alternatives, including several later writings, because it is an especially clear and concise statement of Kohlberg's position.[1]

Drawing on the work of John Dewey and Jean Piaget, Kohlberg posits six stages of moral development. Everyone passes through some of the stages, according to Kohlberg, but few individuals get as far as stages five and six, which are distinguished by their emphasis on moral decision making based on principle. Nevertheless, Kohlberg thinks that it is possible for teachers to help students move from their present level of moral development to the next highest stage. Thus his theory is clearly relevant to educational practitioners.

Kohlberg's theory of moral development stresses the principle of justice. Indeed, for Kohlberg "moral principles are ultimately principles of justice." Consequently, he suggests that moral education should be "centered in justice" and that moral growth may be gauged in terms of students' ascending sense of "institutional justice." For Kohlberg, then, "moral education and civic education are much the same thing."

The primacy accorded to justice in Kohlberg's theory is controversial, but his emphasis on civic education is certainly timely, given the apparent revival of interest in that topic among contemporary educators.[2]

In the final selection, Betty Sichel discusses the relationship between moral judgment and action with reference to Kohlberg's developmental theory. She compares Socrates and Kohlberg on "the problem of weakness of will" and concludes that Kohlberg's theory fails to bridge the gap between moral judgment and moral behavior (to account, that is, for the fact that we do not always act in accordance with our moral convictions). In fact, Sichel thinks that Kohlberg avoids the problem and thereby misses an opportunity to demonstrate the practical value of his theory.

Sichel also maintains that Kohlberg confuses "the process of making a moral judgment with justification of a moral judgment," and she criticizes his preoccupation with cognitive development. The latter, Sichel claims, is accompanied by a failure on Kohlberg's part to appreciate the importance of affective considerations in the development of moral sensitivity.

Thus Professor Sichel raises a number of important questions with regard to what has been perhaps the most widely discussed theory of moral education in recent years. In addition, she provides a useful summary of the critical literature on Kohlberg published prior to the appearance of her own article.[3]

NOTES

1. However, readers interested in later developments concerning experimental programs mentioned in this article may wish to consult R. H. Hersh, D. P. Paollito, and J. Reimer, *Promoting Moral Growth: From Piaget to Kohlberg* (New York: Longman, 1979).

2. In this connection, see R. Freeman Butts, *The Revival of Civic Learning: A Rationale for Citizen Education in American Schools* (Bloomington, IN: Phi Delta Kappa Foundation, 1980). See also the special issue of the *Journal of Teacher Education* 34 (November-December, 1983) devoted to "The Civic Education of the American Teacher."
3. For a later listing of writings by and about Kohlberg, see James S. Lemming, *Foundations of Moral Education: An Annotated Bibliography* (Westport, Ct.: Greenwood Press, 1983).

16. Selections from *Values and Teaching*

Louis E. Raths, Merrill Harmin, and Sidney B. Simon

* * *

DISTINGUISHING VALUES AND VALUING

People grow and learn through experiences. Out of experiences may come certain general guides to behavior. These guides tend to give direction to life and may be called *values*. Our values show what we are likely to do with our limited time and energy.

Since we see values as growing from a person's experiences, we would expect that different experiences would give rise to different values and that any one person's values would be modified as those experiences accumulate and change. A person in the Antarctic would not be expected to have the same values as a person in Chicago. And a person who has an important change in awareness or in patterns of experience might be expected to modify his or her values. Values may not be static if a person's relationships to the world are not static. As guides to behavior, values evolve and mature as experiences evolve and mature.

Moreover, because values are a part of everyday living, they operate in very complex circumstances and usually involve more than simple extremes of right and wrong, good or bad, true or false. The conditions under which behavior is guided, in which values work, typically involve conflicting demands, a weighing and a balancing, and finally an action that reflects a multitude of forces. Thus, values seldom function in a pure and abstract form. Complicated judgments are involved and true values are ultimately reflected in the outcome of life as it is finally lived.

Reprinted, by permission, from *Values and Teaching*, 2d ed. (Columbus, Ohio: Charles E. Merrill Publishing Company, copyright © 1978, 1976 by Louis E. Raths, Merrill Harmin, and Sidney B. Simon), pp. 26–35, 54–58, 80–83.

We therefore see values as being constantly related to the experiences that shape them and test them. For any one person, they are not so much hard and fast verities as they are the results of hammering out a style of life in a certain set of surroundings. After a sufficient amount of hammering, certain patterns of evaluating and behaving tend to develop. Certain things are treated as right, desirable, or worthy. These become our values.

In this book, we shall be less concerned with the particular value outcomes of any one person's experiences than we will with the process that is used to obtain those values. Because life is different through time and space, we cannot be certain what experiences any one person will have. We therefore cannot be certain what values, what style of life, would be most suitable for any person. We do, however, have some ideas about what processes might be most effective for obtaining values. These ideas grow from the assumption that whatever values a person obtains should work as effectively as possible to relate that person to his or her inner and outer worlds in a satisfying and intelligent way.

From this assumption comes what we call the *process of valuing*. A look at this process may make clear how we define a value. Unless something satisfies *all* seven of the criteria noted below, we do not call it a value, but rather a "belief" or "attitude" or something other than a value. In other words, for a value to result, all of the following seven requirements must apply. Collectively, they describe the process of valuing.

1. Choosing freely. If something is in fact to guide our lives, whether or not authority is watching, it will probably have to be freely chosen. If there has been coercion, the results are not likely to stay with us too long, especially when we are out of the range of the source of that coercion. It seems that values must be freely selected if they are to be fully valued. Put another way, the more a person feels that a value has been actively and freely selected, the more she is likely to feel that that value is central to herself.

2. Choosing from alternatives. This definition of values is based on choices made by individuals, and obviously, there can be no choice if there are no alternatives from which to choose. Thus, we say that it makes little sense to include something in the value category when the person involved was aware of no options. Likewise, we would say that the more alternatives open to us in a choice situation, the more likely we are to find something we fully value. When we approach an issue by brainstorming possible options, for example, we increase the likelihood that a value will emerge.

3. Choosing after thoughtful consideration of the consequences of each alternative. The selection of an alternative impulsively or thoughtlessly does not lead to values of the type we are defining. For a value to guide a person's life intelligently and meaningfully, we believe it must emerge in a context of understanding. Only when the consequences of each of the alternatives are understood and considered is a choice not impulsive or thoughtless. There is an important cognitive factor here. The more we understand about the consequences that flow from each alternative, the more we can make an informed choice, a choice that flows from our full intelligence. Thus, we prefer to exclude from the term *values* those choices not making a full use of intelligence.

4. Prizings and cherishing. The values we are defining have positive tones. We prize a value, cherish it, esteem it, respect it, hold it dear. We are happy with our values. A choice, even when we have made it freely and thoughtfully, may be a choice we are not happy to make. We may choose to fight in a war, but be sorry that circumstances make that choice reasonable. In our definition, values flow from choices that we are glad to make. We prize and cherish the guides to life that we call values. We judge them positively.

5. Affirming. When we have chosen something freely, after informed consideration of the alternatives, and when we are proud of our choice, glad to be associated with it, we are likely to want to affirm that choice when asked about it. We are willing that others know of our values. We may even be willing to champion them. If on the other hand we are ashamed of our choice, if we would prefer that no one ever knew about it, we would be dealing with something not as positive as a value, but with something else. We prefer, then, to exclude from the term "values" these choices that we are ashamed to affirm to others.

6. Acting upon choices. Where we have a value, we believe it should show up in aspects of our living, in our behavior. We may do some reading about things we value. We may form friendships or join organizations that nourish our values. We may spend money on values. We very likely budget time or energy for them. In short, for a value to be present, life itself must be affected. Nothing can be a value that does not, in fact, give direction to actual living. The person who talks about something but never does anything about it is acting from something other than a value, in our definition.

7. Repeating. Where something reaches the level of a value, it is very likely to influence behavior on a number of occasions in the life of the person who holds it. It will show up in several different situations, at several different times. We would not think of a behavior that appeared only once in a life as representing a value. Values tend to be persistent. They tend to show up as a pattern in a life.

To review this definition, we see values as based on three processes: choosing, prizing, and acting.

- **Choosing:** (1) freely
  (2) from alternatives
  (3) after thoughtful consideration of the consequences of each alternative
- **Prizing:** (4) cherishing, being happy with the choice
  (5) enough to be willing to affirm the choice to others
- **Acting:** (6) or doing something with the choice
  (7) repeatedly, in some pattern of life

Those processes collectively define valuing. Results of this valuing process are called values.

You may want to pause for a moment and apply the seven criteria for a value to one of your hobbies, be it sewing, skiing or hi-fi. Is it prized, freely and thoughtfully chosen from alternatives, acted upon, repeated, and publicly known? If so, it may be said that you *value* that hobby.

VALUE INDICATORS

Obviously not everything is a value, nor need it be. We also have purposes, aspirations, and beliefs that may not meet all seven of those criteria. However, values often do grow from our purposes, aspirations, and beliefs. Let us briefly discuss some things that could indicate the presence of a value but that are different from values. We call these expressions which approach values, but which may not meet all of the criteria, *value indicators*.

1. Goals or purposes. Purposes give direction to life. If the purpose is important to us, we cherish it and we organize our life to achieve the purpose. This does not mean that every stated purpose is a value. Instead, we should think of a stated purpose as a potential value or a value indicator. If, in our presence, a child states a purpose, we must asume it is merely a stated purpose, until we inquire further and we have an opportunity to pursue with the child whether or not he prizes it, has freely chosen it, has wanted it for some time, and is willing to do what is necessary to achieve it. Some stated purposes are dropped when these processes are applied. The child finds out that what he said is not what he really wants. He might have had what amounts to a passing interest in the idea, but even brief examination often results in depreciation of the stated purpose. Thus a purpose *may* be a value, but, on the other hand, it may not be.

2. Aspirations. We sometimes indicate a purpose that is remote in terms of accomplishment. It is not something that we wish or expect to accomplish today or tomorrow, or within a week or sometimes even in a month. The statement of such an aspiration frequently points to the possibility of something that is valued. We do not know if it is truly a value until we have asked questions which relate to the seven criteria which have been mentioned. When the responses are consistent with those criteria, we can say that we have touched a value.

3. Attitudes. Sometimes we give indications that we have a certain value by expressing an attitude. We say that we are *for* something or *against* something. It is not always a sound practice to infer that such a statement represents a value. It is really cherished? Has some consideration been given to alternatives? Does it come up again and again? Is it related to the life activities of the person who expresses it? Unless these criteria are met, it may be just so many words. That is, it may be an attitude and not a value.

4. Interests. Very often you hear people say that they are interested in something. Care should be taken, however, in concluding that this means that a value is present. Very often when we say we are interested, we mean little more than that we would like to talk about it or to listen to someone talk about it, or that we might like to read a little more in that area. It is not a combination of the criteria which have been proposed. It may be a bit more than a passing fancy, but very frequently it does not work out to be a value.

5. Feelings. Our personalities are also expressed through our feelings and through statements about how we feel. Sometimes we feel hurt. Sometimes we feel outraged. On other occasions we are glad, sad, depressed, excited; and we experience dozens of other feelings. We cannot always say that a value is present related to these feelings. Our feelings may be responses which are dissipated by very brief reflection. We should have to ask a number of questions in order to find out if the feeling reflects an underlying value.

6. Beliefs and convictions. When we hear someone state what she believes, it is all too easy to accept the statement as a value. A woman may believe that there should be discrimination with respect to race, but also may be ashamed of that belief. She may not prize holding it. Moreover, upon examination, she may have doubts about the truth or goodness of that belief. It is the examined belief, the cherished belief, the freely chosen belief, and the belief that pervades life that rises to the stature of a value. The verbal statement provides a pointer, but it is only through careful examination that we get to know whether it represents a value.

7. Activities. We sometimes say about a figure in public life "That's what he says, but what does he do?" We seem to be saying that not until a person does something do we have some idea of what he or she values. With values, as with other things, actions speak louder than words. Of course, it is not true that everything we do represents our values. For example, we are pretty sure that going to church does not necessarily mean a commitment to religion. A person may go to church for many reasons. Another person may go often and regularly to bridge parties, all the while wishing that he did not have to go. He may do a certain kind of work every day without having chosen that work or prizing it very much. In other words, just by observing what people do, we are unable to determine whether values are present. We have to know if the individual prizes what he is doing, if he has chosen to do what he is doing, and if it constitutes a pattern in his life. All by themselves, activities do not tell us enough, but they may indicate a value.

8. Worries, problems, obstacles. We hear individuals talk about worries or problems, and we sometimes infer from the context that we know the values that are involved. Here again we may be giving undue importance to verbal statements. If we were to ask questions bearing upon the seven criteria which have been proposed, we might find out that nothing of great importance is involved; that the statement is made only to further conversation. Many of us talk a good deal, and we may mention problems or worries only as ways of entering into a conversation. Examining the worry or the problem may reveal that something *is* deeply prized, that a belief *is* being blocked, that one's life *is* being disturbed. Under such circumstances, we can be more confident with the judgment that a value is involved.

We have discussed briefly eight categories of behavior which have a significant relationship to valuing. There is no implication that other categories of behavior may not be just as important. However, these eight categories—goals and purposes, aspirations, feelings, interests, beliefs and convictions, attitudes, activities, and worries—are often revealed in the classroom. We believe it is important that opportunities for revealing these become a vital part of teaching, for the next step—as will be discussed in later chapters—is for the teacher to help those children who choose to do so to raise these value indicators to the level of values—that is, to the level on which all seven of the valuing processes operate.

We now want to turn our attention to one of the value processes that seems to have particular importance for work with children growing up in this confused and complex world, the process of choosing.

THE CRUCIAL CRITERION OF CHOICE

Because we see choosing as crucial to the process of valuing, it may be useful at this point to expand on the conditions that must exist if a choice is really to be made.

The act of choosing requires alternatives. We choose something from a group of things. If there is only one possibility, we cannot make a choice, and according to our definition, no value is indicated in that area. Yet so often we eliminate all but one alternative for children or we restrict alternatives available to them. For example, we say to a child that she must either sit silently in her chair or stay after school. Or we ask, "Wouldn't you like to learn your multiplication table, Iris?" If we restrict alternatives, so that the child's preferred choice is not among them, we cannot say that the subsequent choice represents a value. It is common practice to give children either-or choices, both of which may be undesirable from their standpoint, then wonder why they do not value their own behavior.

Unless we open up decisions and include alternatives that a child might really prefer, we may only give the illusion of choice, at least in terms of this value theory. Values must grow from thoughtful, prized choices made from sufficient alternatives.

Thus, when we take the lead in presenting alternatives to children, we should also take some care in seeing to it that the *alternatives have meaning* for them. A student may be familiar with only one of the possible choices and may not be aware of what the others involve. We can discover this by asking questions; and if the student does not know what other choices could mean, we can help the child to understand them better. It is useless, therefore, to ask a young child to make choices from among alternatives he does not understand. It is also misleading to think a choice has been made when, for example, a child selects democracy over autocracy without much understanding of either.

We should also help children to see the probable *consequences of a choice* and find out if they are willing to accept the consequences which may follow. Once a child is on record as being willing to take the consequences, the situation has to that extent been clarified, and the choice has some meaning. Without such an understanding of and acceptance of the consequences, we can hardly call the choice meaningful.

We must further recognize that the child needs to be really *free to choose*. If for many reasons we do not want a child to choose a particular alternative, like setting fire to the house, we should let her or him know that this is not within the realm of choice. We should not try to fool her into thinking that she is a free agent and then disappoint her when we refuse to honor her choice. Rather, we should be clear and forceful when we deny choices. Otherwise, we subvert the faith of the child in the very process of deciding and choosing. When we are concerned with values, we must be willing to give the child freedom to choose.

In short, we are saying that a coerced choice is no choice at all. It is not likely that values will evolve from a choice imbued with threat or bribery, for example. One important implication of this condition of value theory for teachers is a reduction of the punishment and reward systems so widely used in schools. Choices cannot be considered sufficiently free if each one is first to be weighed, approved, disapproved, or graded by someone in charge.

Does this mean that the whole world should be open for choice and that we should respect a child's choice no matter what it is? Most teachers, in the light of policies suggested by a school board or school administrators or out of their own basic needs for order, make quite clear to students that there are areas where choice is not possible. One of these relates to life itself: We do not allow children to engage in activities which might result in serious danger. We say to children about areas where serious injury is a strong possibility that they cannot choose, that their behavior is restricted. The reason we give is that the consequences of an unwise choice are not tolerable or that the alternatives may not be well enough understood to allow a meaningful choice.

Many of us are also against vulgarity in many forms. Profane language, obscene behavior, filth, and dirtiness are matters on which many of us take a stand. We may indicate to children that they may not engage in behavior of this kind; that the policies for which we stand do not allow such things to go on; that deviations from our standards bother us too much to be tolerated. When an individual reveals this kind of behavior, we may intervene directly and perhaps talk privately with the student.

In almost every culture and in many communities there are areas which are sometimes called "hot" or "very delicate." In some communities, it may be matters pertaining to religious issues; in other communities it may be matters relating to political issues. Sex is frequently such an issue. Where these hot issues are matters of civil rights, rights which belong to all individuals, teachers may wish to challenge restrictions or taboos. It is good policy, however, to do this on a professional level, working among colleagues first, and not to take up issues with children in a public way which may be against public policy. It is usually wiser to make attempts to clarify public policy and to modify it before involving the children in affairs that might be extremely embarrassing and extremely provoking. Whether children should be encouraged to reflect and choose in a controversial area is not the point. They *must* reflect and choose if values are to emerge. The question is what teachers should do first in communities which frown upon opening certain issues.

In summary, if children—or adults, for that matter—are to develop values, they must develop them out of personal choices. These choices, if they are possibly to lead to values, must involve alternatives which (1) include ones that are prized by the chooser; (2) have meaning to the chooser, as when the consequences of each are clearly understood; and (3) are freely available for selection.

THE PERSONAL NATURE OF VALUES

We have said that values are a product of personal experiences. They are not just a matter of true or false. We do not go to an encyclopedia or to a textbook for values. Our definition of valuing shows why this is. People have to prize for themselves, choose for themselves, integrate choices into the pattern of their own lives. Information as such does not convey this quality of values. Values emerge from the flux of life itself.

Consequently, we are dealing with an area that is not a matter of proof or consensus, but a matter of experience. If a child says that he likes something, it does not seem appropriate for an older person to say, "You shouldn't like that." Or, if another child should say "I am interested in that," it does not seem quite right for an older person to say to her, "You shouldn't be interested in things like that." If these interests have grown out of a child's experience, they are consistent with his or her life. When we ask children to deny their own lives, we are in effect asking them to be hypocrites. We seem to be saying in an indirect way, "Yes, this is what your life has taught you, but you shouldn't say so. You should pretend that you had a different life." What are we doing to children when we put them into positions like this? Are we helping them to develop values or are we in effect saying that life is a fraud and that a person should learn to live a lie very early in life.

We have an alternative approach to values which will be presented in the following chapters. For now, it is important to note that our definition of values and valuing leads to a conception of these words that is highly personal. It follows that if we are to respect a person's life, we must respect his experience and his right to examine it for values.

It should be expected that in society like ours, governed by our Constitution, teachers might well see themselves as obliged to support the idea that all people are entitled to the views that they have and to the values that they hold, especially where these have been examined and affirmed. Is this not the cornerstone of a free society? As teachers, then, we need to be clear that we cannot dictate to children what their values should be since we cannot also dictate what their environments will be and what experiences they will have. We may be authoritative in those areas that deal with truth and falsity: Do flowers consume oxygen? Where does the state get its income? But where the question involves a personal activity (Do you ever take care of flowers?), an attitude (Are you happy with the way the state spends its money?) or a worry, interest, feeling, purpose, aspiration, or belief, our view is that it is unreasonable for a teacher to assume he or she has the correct answer. By our definition, and as we see it, by social right, values are personal things.

As a matter of cold fact, in the great majority of instances we really do not know what values a child has. We are apt to make inferences which go beyond the available data and to attribute values to children which they do not hold. We probably are better off assuming only that we really do not know. If we are interested in knowing, we might well initiate a process of investigation and inquiry. More attention will be given to this notion later.

LIFE AS A PROCESS

One final analogy will reemphasize our focus. Some people, when they travel, seem to be much more interested in the motels or hotels at which they stop than in the experiences which they have along the way. Other people are much more interested in the road than in the inn. Our approach to values could be likened to the latter group's interest. We are interested in the processes that go on in valuing. We are not much interested in identifying the values which children ultimately hold as a result of these processes. We are more interested in the process because we believe that in a world that changes as rapidly as ours, children must develop habits of examining personal aspirations, purposes, attitudes, feelings, activities, interests, beliefs, and worries if they are to find satisfying ways of integrating their own thoughts, emotions, and behaviors within themselves and in relation to the world.

The development of values is a personal and life-long process. It is not something that is completed by early adulthood. As the world changes, as we change, and as we strive to change the world again, we have many decisions to make and we should be learning how to make these decisions. We should be learning how to value. It is this process that we believe needs to be carried on in the classrooms; it is at least partly through this process that we think children will learn about themselves and about how to make some sense out of the buzzing confusion of the society around them.

* * *

THE CLARIFYING RESPONSE

The most flexible of the value-clarifying strategies is called the clarifying response. It is a response a teacher makes to something a student has said or done when the purpose is to encourage that student to do some extra thinking. As you will see, a clarifying response is sometimes made orally, in an informal dialogue with one student or in the midst of a whole-class discussion. And the clarifying response is sometimes used when writing comments on student papers.

Imagine a student on the way out of class who says, "Ms. Jones, I'm going to Washington, D.C., this weekend with my family." How might a teacher respond? Perhaps, "That's nice," or "Have a good time!" Neither of those responses is likely to stimulate clarifying thought on the part of the student. Consider a different kind of response, for example: "Going to Washington, are you, Sally? Are you glad you're going?"

To sense the clarifying power in that response, imagine the student saying, "No, come to think of it, I'm not glad I'm going. I'd rather play in the Little League game." If the teacher were to say nothing other than "Well, we'll see you Monday," or some noncommittal equivalent, we still might say that the student would have become a little more aware of her life. In this case, she would have recognized that she is capable of doing things that she is not happy about doing. This is not a very big step, and it might be no step at all, but it might contribute to her considering a bit more seriously how much of her life she wants involved in activities that she does not prize or cherish. We say this is a step toward value clarity.

Consider this example. A student says that he is planning to go to college after high school. A teacher who replies, "Good for you," or "Which college?" or "Well, I hope you make it," is probably going to serve purposes other than value clarity. But were the teacher to respond, "Have you considered any alternatives?" the goal of value clarity may well be advanced.

That "alternatives" response is likely to stimulate thinking about the issue, and if the student decides to go to college, that decision is likely to be closer to representing a value than it was before. It may contribute a little toward moving a potential college student from the position of going because "it's the thing to do" to going because he wants to get something out of it.

Here are two other samples of exchanges using clarifying responses.

Student: I believe that all men are created equal.

Teacher: What do you mean by that?

Student: I guess I mean that all people are equally good and none should have advantages over others.

Teacher: Does this idea suggest that some changes need to be made in our world, even in this school and this town?

Student: Oh, lots of them. Want me to name some?

Teacher: No, we have to get back to our spelling lesson, but I was just wondering if you were working on any of those changes, actually trying to bring them about.

Student: Not yet, but I may soon.

Teacher: I see. Now, back to the spelling list.

Teacher: Bruce, don't you want to go outside and play on the playground?

Student: I dunno. I suppose so.

Teacher: Is there something that you would rather do?

Student: I dunno. Nothing much.

Teacher: You don't seem much to care, Bruce. Is that right?

Student: I suppose so.

Teacher: And mostly anything we do will be all right with you?

Student: I suppose so. Well, not anything, I guess.

Teacher: Well, Bruce, we had better go out to the playground now with the others. You let me know sometime if you think of something you would like to do.

THE EFFECTIVE CLARIFYING RESPONSE

You may already sense some criteria of an effective clarifying response, that is, a response that encourages someone to look at his or her life and ideas and to think about them. The following are among the essential elements of this response.

1. The clarifying response avoids moralizing, criticizing, giving values, or evaluating. The adult excludes all hints of "good" or "right" or "acceptability," or their opposites in such responses.
2. It puts the responsibility on the students to look at their behavior or ideas and to think and decide for themselves what it is *they* want.
3. A clarifying response also entertains the possibility that the student will *not* look or decide or think. It is permissive and stimulating, but not insistent. The student will often be expected to decline to answer when asked clarifying questions.
4. It does not try to do big things with its small comments. It works more at stimulating thought related to what a person does or says. It aims

at setting a mood. Each clarifying response is only one of many; the effect is cumulative.

5. Clarifying responses are not used for interview purposes. The goal is not to obtain data, but to help students clarify their own ideas and lives if they want to do so.
6. Usually an extended discussion does not result. The purpose is for the student to think, and this is usually done best alone, without the temptation to justify thoughts to an adult. Therefore, a teacher would be wise to carry on only two or three rounds of dialogue and then offer to break off the conversation with some noncommittal but honest phrase, such as "Nice talking to you" or "I see what you mean better now" or "Let's talk about this another time, shall we?" or "Very interesting. Thanks." (Of course, there is no reason why a student who desires to talk more should be turned aside, the teacher's time permitting.)
7. Clarifying responses are often for individuals. A topic in which Clem might need clarification may be of no immediate interest to Priscilla. An issue that is of general concern, of course, may warrant a general clarifying response—say to the whole class—but even here the *individual* must ultimately do the reflecting. Values are personal things. The teacher often responds to one individual, although others may be listening.
8. The teacher does not respond to everything everyone says or does in a classroom. There are other responsibilities. (In chapter 8, we discuss how clarifying responses can be used with those students who need them most.)
9. Clarifying responses operate in situations where there are no "right" answers—as in situations involving feelings, attitudes, beliefs, or purposes. They are *not* appropriate for drawing a student toward a predetermined answer. The are *not* questions for which the teacher has an answer already in mind.
10. Clarifying responses are not mechanical things that carefully follow a formula. They must be used creatively and with insight, but with their purpose in mind. When a response helps a student to clarify thinking or behavior, it is considered effective.

The ten conditions listed above are very difficult to fulfill for the teacher who has not practiced them. A tendency to use responses to students for the purpose of molding students' thinking is very well established in most teachers' minds. That a function of the teacher is to help the child clarify some of the existing confusion and ambiguity is an unfamiliar idea for many of us. After all, most of us became teachers because we wanted to *teach* somebody something. Most of us are all too ready to sell our intellectual wares. The clarifying strategy requires a different orientation—not that of adding to the child's ideas but rather one of stimulating the child to clarify the ideas he or she already has.

Here is another classroom incident illustrating a teacher using clarifying responses, in this case to help a student see that free, thoughtful choices can be made. The situation is a classroom discussion in which a boy has just made it clear that he is a liberal in his political viewpoints.

Teacher: You say, Glenn, that you are a liberal in political matters?
Glenn: Yes, I am.
Teacher: Where did your ideas come from?
Glenn: Well, my parents I guess, mostly.
Teacher: Are you familiar with other positions?
Glenn: Well, sort of.
Teacher: I see, Glenn. Now, class, getting back to the homework for today . . . (returning to the general lesson).

Here is another actual situation. In this incident, the clarifying response prods the student to clarify her thinking and to examine her behavior to see if it is consistent with her ideas. It is between lessons and a student has just told a teacher that science is her favorite subject.

Teacher: What exactly do you like about science?
Student: Specifically? Let me see. Gosh, I'm not sure. I guess I just like it in general.
Teacher: Do you do anything outside of school to have fun with science?
Student: No, not really.
Teacher: Thank you, Lise. I must get back to work now.

Notice the brevity of the exchanges. Sometimes we call these exchanges "one-legged conferences," because they often take place while a teacher is on one leg, pausing briefly on the way elsewhere. An extended series of probes might give the student the feeling of being cross-examined and might make the child defensive. Besides extended discussion would give the student too much to think about. The idea is, to raise a few questions, without moralizing, leave them hanging in the air, and then move on. The student to whom the questions are addressed and other students who might overhear may well ponder the questions later, in a lull in the day or in the quiet moments before falling asleep at night. These are gentle prods, but the effect is to stimulate a student who is ready for it to choose, prize, and act in ways outlined by our theory. And, as the research reported in chapter 12 demonstrates, these one-legged conferences add up to large differences in some students' lives.

* * *

IN CONCLUSION

Basic to the approach to values offered in this book is the clarifying response. The clarifying response is usually aimed at one student at a time, often through brief, informal conversations held in class, in hallways, on the playground, or any place else where the teacher comes in contact with a student who does or says something to trigger such a response. Clarifying responses are also used during whole-class discussions and as comments written in the margins of students' papers.

Especially ripe for clarifying responses are such things as expressions of students' feelings, beliefs, convictions, worries, attitudes, aspirations, purposes, interests, problems, and activities. These sometimes indicate a value or a potential value and thus we refer to them as *value indicators.* A teacher who is sensitive to these expressions finds many occasions for useful clarifying responses.

The purpose of the clarifying response is to raise questions in the students' minds, to prod them gently to examine their lives, their actions, and their ideas, with the expectation that some will want to use this prodding as an opportunity to clarify their understandings, purposes, feelings, aspirations, attitudes, beliefs, and so on.

Participating in such exchanges, some students may choose the thoughtful consistency between words and deeds that characterizes values. But not everything need be a value. Beliefs, problems, attitudes, and all the rest are parts of life too.

It may be useful to list some things that a clarifying response is *not.*

1. Clarifying is not therapy.
2. Clarifying is not used on students with serious emotional problems.
3. Clarifying is not a single one-shot effort, but depends on a program consistently applied over a period of time.
4. Clarifying avoids moralizing, preaching, indoctrinating, inculcating, or dogmatizing.
5. Clarifying is not an interview, nor is it meant to be done in a formal or mechanical manner.
6. Clarifying is not meant to replace the teacher's other educational functions.

If clarifying is none of the above, what is it? It is an honest attempt to help students look at their lives and to encourage them to think about life, and to think about it in an atmosphere in which positive acceptance exists. No eyebrows are raised. When a student reveals something before the whole class, he or she must be protected from snickers from other class members. It is essential to maintain an environment where searching is highly regarded.

We emphasize that students will probably not enter the perplexing process of clarifying value for themselves if they perceive that the teacher does not respect them. If trust is not communicated—and the senses of students for such matters can be mystifyingly keen—students may well play the game, pretending to clarify and think and choose and prize, while being as unaffected as by a tiresome morality lecture. This is a difficult and important problem, for it is not easy to be certain you are communicating trust, whether or not you believe you are doing so. (A moot point is whether some people can communicate a trust that they, in fact, do not have.) One must be chary about concluding that a teacher who says the right words is getting the results desired. There is a spirit, a mood, required that we cannot satisfactorily describe or measure except to say that it seems related to a basic and honest respect for students. It may be fair to say that a teacher who does not communicate this quality will probably obtain only partial results.

For many teachers a mild revolution in their classroom methodology will be demanded if they are to do very much with the clarification of values. For one thing, they will have to do much less talking and listen that much more, and they will have to ask different kinds of questions from the ones they have asked in past years. Teachers usually favor questions that have answers that can be scaled from "right" to "wrong." No such scoring can be applied to answers to clarifying questions.

The rewards for giving up the old patterns may not come right away, but there is mounting evidence that teachers who act responsively begin to have small miracles happening in their classrooms. They often see attendance go up, grades rise, and interest and excitement in learning crackle. They witness students who had been classified as apathetic, listless, and indifferent begin to change. In the words of one teacher, "students get their heads off their elbows and use those elbows to wave hands in the air."

In brief, one might see the clarifying response fitting into the value-clarifying method in the following way.

1. First, look and listen for value indicators, statements or actions which suggest that there could be a value issue involved.

   It is usually wise to pay special attention to students who seem to have particularly unclear values. Note especially children who seem to be very apathetic or indecisive, who seem to be very flighty, or who drift from here to there without much reason. Note, also, children who overconform, who are very inconsistent, or who play-act much of the time.

2. Secondly, keep in mind the goal: children who have clear, personal values. The goal therefore requires opportunities for children to use the processes of (a) choosing freely, (b) choosing from alternatives, (c) choosing thoughtfully, (d) prizing and cherishing, (e) affirming, (f) acting upon choices, and (g) examining patterns of living. We provide these opportunities with the expectation that the results of these processes are better understandings of what a person stands for and believes in and more intelligent living.
3. Thirdly, respond to a value indicator with a clarifying question or comment. This response is designed to help the student use one or more of the seven valuing processes listed above. For example, if you guess that a child does not give much consideration to what is important to *him,* you might try a clarifying response that gets at prizing and cherishing. Or the form of the value indicator may suggest the form of the clarifying response. For example, a thoughtless choice suggests responses that get at choosing, and a fine-sounding verbalization suggests responses that get at incorporating choices into behavior.

* * *

17. Clarifying Values Clarification: A Critique

John S. Stewart

Much has been written about the positive claims of Values Clarification,[a] now a well-developed educational fad. The acceptance of this approach to values education has been truly phenomenal. Paradoxically and ironically, this acceptance has been, for the most part, somewhat unreflective, uncritical, and unclarified. In fact, to point out the flaws, discuss the inconsistencies, or challenge the claims or validity of Values Clarification to teachers is frequently tantamount to attacking their children. Thus VC has become almost a sacred cow. Of late, however, I have observed an increasing willingness on the part of teachers to begin to face up to some of the inadequacies and problems of this approach. I find this especially true with teachers who have been to one or more VC workshops, have made extensive efforts to implement the ideas, and have experienced some success. The most frequent comment I hear is, "But what do we do now?" or "Where do we go from here?" These teachers will tell you that VC is good, but it simply isn't enough, and eventually both the teachers and the students become bored.

Perhaps a good place to begin a critical look is to deal with the superficiality of Values Clarification. Why isn't it enough? Because, I believe, it deals primarily with the *content* of values and somewhat with the *process* of valuing, but ignores the most important aspect of the issue—namely, the *structure* of values and valuing, especially structural development. To oversimplify a very complex subject, the content of values/moral judgments is the "what" that is expressed by the person, the answer to a question or a dilemma; whereas the structure is the underlying cognitive logic on which the content is based, or the "why" that generated the answer.

Reprinted, by permission, from the *Phi Delta Kappan* 56 (June 1975): 684–688.

The term *Values Clarification* will be capitalized throughout to show its "patented" nature and to indicate that it is only one possible approach to clarifying values, much as a Kodak is only one kind of camera. The initials VC will also be used to stand for Values Clarification.

Jean Piaget, Lawrence Kohlberg, and other structural-developmental researchers have discovered that there are universal patterns of logic and principles of justice that form the structure of intelligence, including moral judgment. These cognitive structures are *constructed* through the transactions between the individual and the world, and develop through an invariant sequence of stages. Piaget has identified four major periods or stages in the development of intelligence, and Kohlberg has identified six stages in the development of moral judgment. Kohlberg's extensive longitudinal research has shown that the structure of values/moral development is considerably more important than the content of particular judgments. Yet it is the content of particular value judgments—the most superficial and ephemeral part of this complex aspect of human existence—on which the methodology of Values Clarification focuses. This focus leads Louis E. Raths, Sidney B. Simon, and Merrill Harmin to conclude that "a person in the Antarctic would not be expected to have the same values as a person in Chicago."[1] True, if we are dealing with the content of values, which is, of course, relative to culture. But not true, as the structuralists have been able to discover, if we are dealing with the structural aspects, which are universal to all people and cultures. People in the Antarctic, or anyplace else, for that matter, have available to them the same logic and principles of justice and pass through the same developmental stages as do people in the United States. But the issue of relativity is not the point here; that will be discussed shortly. The point is that much of what is done in Values Clarification deals with content rather than with structure. A survey of the VC strategies and methods reveals the content focus, and also reveals the superficiality, banality, and triviality of a great deal of the questions, issues, and activities VC deals with.

Considerably more important than VC's superficiality is its reliance on peer pressure and a tendency toward coercion to the mean in many activities. In spite of the emphasis on individuality and the many statements in the VC literature about avoiding peer pressures, many of the strategies and the social aspects of the methodology are highly conducive to peer pressure, especially among highly sensitive teen-agers and even adults who are particularly attuned to the judgments of others. The emphasis on frequent public affirmation of positions, for example, carries this danger. Given the social dynamics of a normal group of teen-agers, only the most popular or the strongest dare express their honest opinions and feelings about values/moral issues publicly without fear of ridicule or rejection.

Space permits only one example, but it easily communicates the problem. One of the most frequently used strategies is the "Values Continuum,"[2] which involves having students take positions on issues presented on a continuum from one extreme to its opposite. One of the items in this strategy asks, "How do you feel about premarital sex?" The two ends of the continuum are 1) Virginal Virginia (sometimes called Gloves Gladys) and 2) Mattress Millie. Virginal Virginia "wears white gloves on every date" and

Mattress Millie "wears a mattress strapped to her back." Now consider the very shy, sensitive, and fearful girl in the class as an extreme example—the girl who's tremendously concerned about her standing with the other girls, or the boys, or the teacher. Suppose that her position on this issue is clear, even as the result of having applied the principles of Values Clarification, and that she truly believes in either one of the two extreme positions. Would she be likely to affirm publicly such a position in this situation? I would think not. The risks would simply be too great. Assuming that she also does not want to run the risk of being judged somehow for passing, one of the legitimate choices in all VC exercises, she might be inclined to express a middle position.

This is what I mean by coercion to the mean, and I see it as a great factor in many of the VC strategies, especially those strategies like the Values Continuum in which the extreme positions are so value-specific and/or emotionally loaded as to preclude them as legitimate alternatives for public affirmation for many people. Even the middle choice in many cases is equally unacceptable, because of the implications of its wording, e.g., "compulsive moderate." Some of the items in the Values Continuum offer extreme choices in pejorative terms along with compulsive moderate as a middle choice. What is one to choose in cases like this? Again, given the dynamics of teen-age social relationships, the Values Clarification approach can be harmful, or at least can actually lead to anything but true clarification. The VC creators include many forewarnings about some of these problems, yet tend to minimize them, offer unrealistic and simplistic solutions, or be glib about them. In fact, both in the design of many of the strategies and in the directions given for their use, the biases (values) of the authors are blatantly obvious and enormously influence the way people respond.

A related criticism deals even more directly with the judgmental nature of some aspects of Values Clarification methodology, in spite of the enormous emphasis the VC authors place on the necessity for being nonjudgmental in clarifying values. One of the so-called "clarifying responses" recommended is worded, "Do you do anything about that idea?"[3] The authors say about this, "A verbalization that is not lived has little import and is certainly not a value." This is one of their basic tenets, but I find it highly judgmental, and considering the limitations of life that drastically reduce our opportunities to act on all of our ideas, it may be absurd. But the mere challenge, no matter how ostensibly neutral or nonjudgmental the question asked, carries an implicit threat or implies a judgment. The VC position on this point is an oversimplified generalization, and is also highly moralistic in nature. (And moralizing is even more strongly rejected by the VC creators.) In order to evaluate (not judge) a person's acting or not acting on any given issue, one must know a lot more about the situation, the person's developmental progress, the costs or dangers involved in acting or not acting, and many other factors. But the judgmental nature of Values Clarification is pervasive. The creators have built a methodology based on their own values, which are frequently in conflict. They claim value neutrality with regard to the content of the methodology, but fail to see that their own values are built into the methodology. As a result, they fail to see how enormously judgmental are many of their questions, techniques, and strategies.

The demand for public affirmation and action which is such an important part of the VC philosophy, in addition to being highly judgmental, is also potentially dangerous, especially when one is working with teen-agers. Research conducted by social psychologists has revealed that when people take public positions or are forced to act they tend to cling to the beliefs or values involved, even if those beliefs or values are tentative or not genuinely held at the time of the commitment or action. Once the stand has been taken and the action completed, there is a tendency to live with it and not risk embarrassment or threat by changing later. To go back on a public affirmation or action is often to lose face. Premature affirmation or action, therefore, can be a very dangerous thing to induce. I have seen many teen-agers make premature statements or perform actions they later regretted, but felt obliged to maintain until even they came to believe in what they had said or done. During the important developmental years of adolescence and youth, there is a need for genuine commitment, rational action, and public affirmation. But the risks are also great, and such behaviors should not be artificially induced or prematurely generated.

The Values Clarification position on action seems to be one-dimensional and stereotypical. Throughout their writings, VC authors dogmatically assert that what is not acted upon is not a value. Action is one of the seven criteria that must be met for the idea to qualify as a value. In fact, mere action is not sufficient; it must be repeated action. Of the VC writers, only Howard Kirschenbaum acknowledges some of the absurdities in this position.[4] What constitutes action? How many times must one act in order for it to meet the seventh and final criterion of "acting repeatedly, in some pattern of life"?[5] This assertion ignores the realities of life and limits most people to a very small number of values. Only martyrs are able to affirm and act consistently and publicly on some of the highest values. Would the Values Clarification authors be willing to claim that only Martin Luther King and others who were able, for various reasons, to act repeatedly on their beliefs and values about racial justice could claim racial equality as a value? If so, then they are claiming courage and a whole host of other highly questionable factors as criteria for holding values.

Of the criticisms made against Values Clarification, probably none is made more frequently or more loudly than the charge that it is inadequate, ineffective, and possibly even dangerous because of its basic moral relativism. But to see Values Clarification as morally or ethically relativistic is to see only half of its very confused philosophy. It purports to believe, on the one hand, that values are personal, situational, individually derived, not amenable to objective evaluation, and relative. It offers, on the other hand, a list of values-based behaviors that are undesirable and objectively inferior and can be cured with Values Clarification. This list, mentioned frequently in the literature, contains the following: apathy, flightiness, uncertainty, inconsistency, drifting, overconforming, overdissenting, and role playing. Counterbalancing this list of "vices," of course, would be the corresponding and opposite "virtues." The Values Clarification "bag of virtues," to use Kohlberg's term, is augmented by Simon and Polly deSherbinin in their article in this *Kappan*, viz., purposefulness, productivity, strong beliefs, thoughtfulness, consideration, zestfulness, manageability, and some others.[6] A consideration of some of the VC writers' own statements will reveal the confusion and conflict. Raths, Harmin, and Simon say:

> The point has been made that our values tend to be products of our experiences. They are not just a matter of true or false. One cannot go to an encyclopedia or to a textbook for values. The definition that has been given makes this clear. One has to prize for himself, choose for himself, integrate choices into the pattern of his own life. Information as such doesn't convey this quality of values. Values come out of the flux of life itself.
>
> This means that we are dealing with an area that isn't a matter of proof or consensus. It is a matter of experience.[7]

On the following page, they make this important statement:

> As teachers, then, we need to be clear that we cannot dictate to children what their values should be, since we cannot also dictate what their environments should be and what experiences they will have. *We may be authoritative in those areas that deal with truth and falsity.* (Italics added)

There are a few problems here that merit discussion. First, both of the above statements make it clear that the authors see values as relative. But the italicized sentence in the second statement completely contradicts the basic relativity premise of Values Clarification. What are those areas of truth and falsity about which we may be authoritative? And on what basis may we be authoritative? What are the criteria for truth and falsity? Why are these areas, whatever they are, not subject to the same conditions as other areas of experience? And is this statement itself not a values statement of the most absolute order? It is interesting that this most important statement is made in the stream of thought about the relativity and personal nature of values without any discussion, qualification, or elucidation of its meaning.

Some light is shed on the above problem by an examination of other VC writings. It is especially interesting to note that the two statements about to be considered are from articles addressed to religious educators. In "Three Ways To Teach Church School," Simon clearly enunciates the basic VC position: "Values are very complex and very personal; there are no 'right' values."[8] In "Value Clarification: New Mission for Religious Education," Simon and two other writers say:

> But aren't there any absolute values that we can teach our students? Many of our readers undoubtedly ask this question.
>
> We would allow that *there are certain absolutes* essential to Christian education: belief in God, the Resurrection of Christ, the Holy Spirit, and the veracity of the Ten Commandments, to name a few. *But there are really far fewer immutable and absolute values than most of us realize. What students need most is not a list of values of today but a process for finding new expressions of values that are yet to come. This is what value training is all about.*[9] (Italics added)

Perhaps this makes clear why I have classified Values Clarification as belonging to an ethical position and form of values/moral education I have labeled *absolute relativism.*[10] The basic Values Clarification premise is clear: *All* values statements are relative—*except* 1) this one, 2) those that are essential for the Values Clarification theory and methodology, and 3) those deemed absolute by groups or organizations who want to use Values Clarification but keep their own values systems intact, e.g., Christian educators, schools, and others. In other words, values are absolutely relative, or perhaps they are relatively absolute. In one article, religious educators are told "there are no 'right' values." In a different article, they are told by the same author that there are some "right" values—"right," that is, if Christians say they are right, but there are fewer of these "immutable and absolute values than most of us realize." Obviously, this ambiguous and self-contradictory position would have to apply to *any* person, group, or organization. Or the Values Clarification people would have to decide for whom it does or does not apply—which would, of course, make them the absolute judges, a position they do not want to be in—or do they? This unresolved problem in VC shows us how far the old absolute/relative argument can take us. But, it seems to me, the VC leaders need to wrestle with the problem and make a clear and forthright statement of their position—a public affirmation of their values, that is.

It is interesting to note that, in a published exchange of opinion between Kohlberg and Simon, Simon wondered why Kohlberg thought he conveyed such "overwhelming relativism." He added, "But because he believes it so strongly, I'd like not to just write it off."[11]

Many of the problems presented so far stem from one basic problem: the failure of the creators and leaders of this movement to develop—thoroughly, systematically, and continuously—an integrated conceptual framework, a theory. Except for some of the questions raised by Kirschenbaum and some of his attempts to reformulate the theory, Values Clairification has accepted, and has remained committed to, an inadequate theory it inherited from Louis Raths—a theory that is philosophically indefensible and psychologically inadequate. To claim, as do Raths and Simon, that the theory is derived from John Dewey's ideas, especially his 1939 *Theory of Valuation*, may be somewhat of an exaggeration. It possibly reflects a misunderstanding of Dewey's thought, and certainly it needs explication not found in the VC literature. Some of Dewey's ideas have been adopted, but unsystematically and without regard for the context or the larger system of Dewey's thought. One significant difference between Dewey and the VC people is that his understanding of democracy was quite different from theirs. They use the same word but mean somewhat different things. Also, Dewey was a contextualist, but he was not an absolute relativist. Furthermore, and of primary importance, the Values Clarification people have failed to consider adequately or come to grips with either social psychology in general or developmental psychology in particular.

The research offered to support the validity and efficacy of Values Clarification is weak and seriously flawed, especially the research cited in Chapter 10 of the basic VC text, *Values and Teaching*, by Raths, Harmin, and Simon. Many of their statements are nothing more than the subjective claims of the teachers and/or the researchers involved in the studies, and the objective findings are frequently inconclusive. Moreover, the research designs have so many problems that even the conclusive findings cannot be taken seriously. Admittedly, to conduct research in this field is extremely difficult, but to recognize this would be better than to cite dubious findings and defective designs as research support for the methodology. One of the most serious problems facing researchers in this arena is to control for the Hawthorne effect. Many of the observed outcomes of Values Clarification may be the result of the increased attention and/or the overall warming of the environment that may be a natural concomitant of this type of activity or of any change from one approach to another. It may be that any, or almost any, humane, open, and more affective approach could generate similar results. But this is an open question that requires a great deal of attention. In the meantine, the Values Clarification people need to be extremely conservative in their claims for empirical support.

The Simon and deSherbinin article in this *Kappan* is, in my opinion, unfortunately glib, trite, and superficial, as well as unnecessarily evangelistic. Furthermore, it reveals little about Values Clarification theory, methodology, or progress. There are a number of points worth making, however, with regard to this statement. First, the pervasive theoretical confusion alluded to earlier is manifested in the opening lines of the article, which support the claim that VC writers have failed to conceptualize values, valuing, and other fundamental concepts adequately in their framework. The opening paragraph of the article says: "There's no place to hide from your values. *Everything you do reflects them.* Even denying that *your values show in your every act* is a value indicator" (italics added).

In previous writings, however, VC writers have strongly emphasized that values are extremely rare phenomena reflected and manifested in very specific behaviors that fall under the umbrella defined by the now well-known seven criteria: *choosing*—1) freely, 2) from alternatives, 3) after reflection; *prizing*—4) cherishing, being happy with the choice, 5) willing to affirm the choice publicly; *acting*—6) doing something with the choice, 7) repeatedly, in some pattern of life. Raths, Harmin, and Simon assert emphatically: "Unless something satisfies *all* seven of the criteria . . . , we do not call it a value" (italics in original).[12] And in the same book (p. 32), they make the following statement that directly contradicts the quotation cited above:

> Of course, *it isn't true that everything we do represents our values. . . . Just by observing what people do, we are unable to determine if values are present.* We have to know if the individual prizes what he is doing, if he has chosen to do what he is doing, and if it constitutes a pattern in his life, etc. (Italics added)

Now it is precisely this kind of conflict, ambiguity, and poor scholarship that I find characteristic of the theory offered as the foundation for Values Clarification.

Another need not met by this recent article is for an assessment of the current status of the Values Clarification position and framework. There is no indication whatsoever by Simon and deSherbinin that they have given any thought to the problem, nor is there any mention of the work of their own colleague, Howard Kirschenbaum, in this direction. Are we safe in concluding that the authors of this article see no problems with the theory and feel no need to respond to Kirschenbaum? Are we safe in concluding that these authors feel no need to respond to the frequent and serious criticisms of the relativistic position of Values Clarification? The indications are that they seem to be content with the original theory.

The hidden agenda, or bag of virtues, of Values Clarification—or at least of Simon and deSherbinin—is evident in many places. You should be "purposeful," "know what you want," and "don't fritter away your time." Children who have benefited from Values Clarification, according to research cited, "developed a favorable attitude toward learning" and "became far less noisy and generally recalcitrant." "Productivity" is definitely on the list. A number of virtues in the bag are readily identifiable from their statement about how Values Clarification improves critical thinking:

> You can find people who've been in values clarification for some time who can see through other people's foolishness. They're not taken in by smooth talkers. They seem to get a large picture of what's good and beautiful and right, and to know what's wrong. They're less vulnerable to fads and to hopping on the latest bandwagon of new things to own, to do, and to believe in.

The writers do not, of course, define what constitutes "other people's foolishness" or being a "smooth talker," nor do they tell us precisely what the benefactors of VC discovered to be "good," "beautiful," "right," or "wrong." But we do learn that "when people know what they want, believe strongly, and follow up on commitments, *they are nicer people to have around*" (italics added). And in the paragraph immediately following the preceding statement, we find that "the nicer people" are also (besides being productive and purposeful) "creative," sharing, warm, consistent, and "thoughtful and considerate students, ones who know how to work on *decent* relationships with each other" (italics added). Later, we find "zestful" added to the list and also discover that people "who work on their values . . . rarely flip-flop around." On the contrary, they "know what they want" and they "know how to ask for it."

Curiously, the virtues and vices that can be extrapolated from this article are strangely reminiscent of those of the old character education approach, many Sunday school teachers, the Boy Scouts and Girl Scouts, and most traditional schools. It is the preaching, moralizing, and manipulating of these approaches, organizations, and institutions that the Values Clarification creators and workers have so effectively and so justifiably criticized. One of the greatest contributions of Values Clarification to the field of values/moral education has been its direct confrontation with traditional-authoritarian education, clearly revealing the inadequacy, inhumanity, and injustice of this approach. For Values Clarification to fall into the same traps, then, seems inexcusable. In fact, even in their article Simon and deSherbinin make the following claim: "Moralizing offers the illusion of looking like the right way to go, but the whole focus of trying to shape and manipulate people into accepting a given set of values is doomed to failure." The claim in this statement relates to the pragmatic inadequacy of moralizing, shaping, and manipulating. But anyone familiar with the VC literature and people knows that their opposition is more than pragmatic. It is made primarily on moral grounds. Yet the examples of VC research and applications I have already cited clearly reveal the particular values sought and the manipulations used to obtain them. I suggest that Simon and deSherbinin make statements they would disdain if read in connection with some other approach to values/moral education. Take, for example, the following statement, made in a plea to go beyond the traditional modeling approach: "Values clarifiers believe, however, that people who go through the process of deciding what they value will in the end reflect the ways one would hope, in any event, all good teachers would behave." How can these authors, in keeping with their basic premises, make a statement like this?

I hope that this critique has pointed out some problems and raised some questions that will generate deeper reflection on the extremely important and influential movement known as Values Clarification. There is no question in my mind that this approach to values/moral education has made some significant contributions. If nothing more, it has increased awareness of the issues and generated an enormous amount of movement away from the restrictive and inhibiting forms of traditional moral education.

My assignment has been to take a look at the problems and weaknesses of Values Clarification. In spite of its significant and positive influence, I believe that VC has some potentially serious, even dangerous, problems and implications. Unfortunately, in my judgment, the movement is rooted in a confused philosophy of absolute relativism and in an inadequate psychology of instrumental individualism. It carries a mixed and conflicting bag of virtues. Some of its methodology is excellent and highly useful for approaching values/moral development; but much of its methodology is faulty and some of it creates an illusion of significance that can lead one to think that more is happening than is really happening.

In conclusion, if Values Clarification is going to fulfill its promise and exceed its past glories, it will have to make some major reformulations and extend itself far beyond its present theoretical and empirical base. More than anything, the Values Clarification leaders must apply their basic principles to their own work: They must clarify Values Clarification in order to make it possible to choose wisely, publicly affirm a more rational program, and act repeatedly with humanity and justice to improve values/moral education.

NOTES

1. Louis E. Raths, Merrill Harmin, and Sidney B. Simon, *Values and Teaching: Working with Values in the Classroom* (Columbus, O.: Charles E. Merrill, 1966).
2. Sidney B. Simon, Leland Howe, and Howard Kirschenbaum, *Values Clarification: A Handbook of Practical Strategies for Teachers and Students* (New York: Hart, 1972).
3. Raths, Harmin, and Simon, op. cit., p. 58.

4. Howard Kirschenbaum, "Beyond Values Clarification," in Howard Kirschenbaum and Sidney B. Simon, eds., *Readings in Values Clarification* (Minneapolis: Winston Press, 1973).
5. Raths, Harmin, and Simon, op. cit., p. 30.
6. See pp. 679–83, this *Kappan*.
7. Raths, Harmin, and Simon, op. cit., p. 36.
8. Sidney B. Simon, "Three Ways To Teach Church School," in Kirschenbaum and Simon, op. cit., pp. 237–40.
9. Sidney B. Simon, Patricia Daitch, and Marie Hartwell, "Value Clarification: New Mission for Religious Education," in Kirschenbaum and Simon, op. cit., pp. 241–46.
10. See John S. Stewart, "Values Development Education," in Ted W. Ward and John S. Stewart, *Final Report: An Evaluative Study of the High-School-Use Films Program of Youth Films, Incorporated* (East Lansing, Mich.: School of Education, Michigan State University, 1973); idem, *Essays on Values Development Education* (unpublished booklet available from the Values Development Education Program, College of Education, 213 Erickson Hall, Michigan State University, East Lansing, Mich., 1974a); and idem, "Toward a Theory for Values Development Education" [unpublished doctoral dissertation, Michigan State University, 1974b (available from University Microfilms, Inc., Ann Arbor, Mich.)].
11. From an exchange of opinion between Kohlberg and Simon which appeared in the December, 1972, issue of *Learning* magazine.
12. Raths, Harmin, and Simon, op. cit., p. 28.

18. The Cognitive-Developmental Approach to Moral Education

Lawrence Kohlberg

In this article, I present an overview of the cognitive-developmental approach to moral education and its research foundations, compare it with other approaches, and report the experimental work my colleagues and I are doing to apply the approach.

I. MORAL STAGES

The cognitive-developmental approach was fully stated for the first time by John Dewey. The approach is called *cognitive* because it recognizes that moral education, like intellectual education, has its basis in stimulating the *active thinking* of the child about moral issues and decisions. It is called developmental because it sees the aims of moral education as movement through moral stages. According to Dewey:

> The aim of education is growth or *development*, both intellectual and moral. Ethical and psychological principles can aid the school in the *greatest of all constructions—the building of a free and powerful character*. Only knowledge of the *order and connection of the stages in psychological development can insure this*. Education is the work of *supplying the conditions* which will enable the psychological functions to mature in the freest and fullest manner.[1]

Dewey postulated three levels of moral development: 1) the *pre-moral* or *preconventional* level "of behavior motivated by biological and social impulses with results for morals," 2) the *conventional* level of behavior "in

Reprinted, by permission, from the *Phi Delta Kappan* 56 (June 1975): 670–677.

which the individual accepts with little critical reflection the standards of his group," and 3) the *autonomous* level of behavior in which "conduct is guided by the individual thinking and judging for himself whether a purpose is good, and does not accept the standard of his group without reflection."[a]

Dewey's thinking about moral stages was theoretical. Building upon his prior studies of cognitive stages, Jean Piaget made the first effort to define stages of moral reasoning in children through actual interviews and through observations of children (in games with rules).[2] Using this interview material, Piaget defined the pre-moral, the conventional, and the autonomous levels as follows: 1) the *pre-moral stage*, where there was no sense of obligation to rules; 2) the *heteronomous stage*, where the right was literal obedience to rules and an equation of obligation with submission to power and punishment (roughly ages 4–8); and 3) the *autonomous stage*, where the purpose and consequences of following rules are considered and obligation is based on reciprocity and exchange (roughly ages 8–12).[b]

In 1955 I started to redefine and validate (through longitudinal and cross-cultural study) the Dewey-Piaget levels and stages. The resulting stages are presented in Table 1.

We claim to have validated the stages defined in Table 1. The notion that stages can be *validated* by longitudinal study implies that stages have definite empirical characteristics.[3] The concept of stages (as used by Piaget and myself) implies the following characteristics:

1. Stages are "structured wholes," or organized systems of thought. Individuals are *consistent* in level of moral judgment.
2. Stages form an *invariant sequence*. Under all conditions except extreme trauma, movement is always forward, never backward. Individuals never skip stages; movement is always to the next stage up.
3. Stages are "hierarchical integrations." Thinking at a higher stage includes or comprehends within it lower-stage thinking. There is a tendency to function at or prefer the highest stage available.

Each of these characteristics has been demonstrated for moral stages. Stages are defined by responses to a set of verbal moral dilemmas classified according to an elaborate scoring scheme. Validating studies include:

1. A 20-year study of 50 Chicago-area boys, middle- and working-class. Initially interviewed at ages 10–16, they have been reinterviewed at three-year intervals thereafter.
2. A small, six-year longitudinal study of Turkish village and city boys of the same age.
3. A variety of other cross-sectional studies in Canada, Britain, Israel, Taiwan, Yucatan, Honduras, and India.

Table 1. Definition of Moral Stages

I. Preconventional level

At this level, the child is responsive to cultural rules and labels of good and bad, right or wrong, but interprets these labels either in terms of the physical or the hedonistic consequences of action (punishment, reward, exchange of favors) or in terms of the physical power of those who enunciate the rules and labels. The level is divided into the following two stages:

Stage 1: *The punishment-and-obedience orientation.* The physical consequences of action determine its goodness or badness, regardless of the human meaning or value of these consequences. Avoidance of punishment and unquestioning deference to power are valued in their own right, not in terms of respect for an underlying moral order supported by punishment and authority (the latter being Stage 4).

Stage 2: *The instrumental-relativist orientation.* Right action consists of that which instrumentally satisfies one's own needs and occasionally the needs of others. Human relations are viewed in terms like those of the marketplace. Elements of fairness, of reciprocity, and of equal sharing are present, but they are always interpreted in a physical, pragmatic way. Reciprocity is a matter of "you scratch my back and I'll scratch yours," not of loyalty, gratitude, or justice.

II. Conventional level

At this level, maintaining the expectations of the individual's family, group, or nation is perceived as valuable in its own right, regardless of immediate and obvious consequences. The attitude is not only one of *conformity* to personal expectations and social order, but of loyalty to it, of actively *maintaining*, supporting, and justifying the order, and of identifying with the persons or group involved in it. At this level, there are the following two stages:

Stage 3: *The interpersonal concordance or "good boy—nice girl" orientation.* Good behavior is that which pleases or helps others and is approved by them. There is much conformity to stereotypical images of what is majority or "natural" behavior. Behavior is frequently judged by intention—"he means well" becomes important for the first time. One earns approval by being "nice."

Stage 4: *The "law and order" orientation.* There is orientation toward authority, fixed rules, and the maintenance of the social order. Right behavior consists of doing one's duty, showing respect for authority, and maintaining the given social order for its own sake.

III. Postconventional, autonomous, or principled level

At this level, there is a clear effort to define moral values and principles that have validity and application apart from the authority of the groups or persons holding these principles and apart from the individual's own identification with these groups. This level also has two stages:

Stage 5: *The social-contract, legalistic orientation,* generally with utilitarian overtones. Right action tends to be defined in terms of general individual rights and standards which have been critically examined and agreed upon by the whole society. There is a clear awareness of the relativism of personal values and opinions and a corresponding emphasis upon procedural rules for reaching consensus. Aside from what is constitutionally and democratically agreed upon, the right is a matter of personal "values" and "opinion." The result is an emphasis upon the "legal point of view," but with an emphasis upon the possibility of changing law in terms of rational considerations of social utility (rather than freezing it in terms of Stage 4 "law and order"). Outside the legal realm, free agreement and contract is the binding element of obligation. This is the "official" morality of the American government and constitution.

Stage 6: *The universal-ethical-principle orientation.* Right is defined by the decision of conscience in accord with self-chosen *ethical principles* appealing to logical comprehensiveness, universality, and consistency. These principles are abstract and ethical (the Golden Rule, the categorical imperative); they are not concrete moral rules like the Ten Commandments. At heart, these are universal principles of *justice,* of the *reciprocity* and *equality* of human *rights,* and of respect for the dignity of human beings as *individual persons* ("From Is to Ought," pp. 164, 165).

—Reprinted from *The Journal of Philosophy,* October 25, 1973

With regard to the structured whole or consistency criterion, we have found that more than 50% of an individual's thinking is always at one stage, with the remainder at the next adjacent stage (which he is leaving or which he is moving into).

With regard to invariant sequence, our longitudinal results have been presented in the *American Journal of Orthopsychiatry* (see footnote 8), and indicate that on every retest individuals were either at the same stage as three years earlier or had moved up. This was true in Turkey as well as in the United States.

With regard to the hierarchical integration criterion, it has been demonstrated that adolescents exposed to written statements at each of the six stages comprehend or correctly put in their own words all statements at or below their own stage but fail to comprehend any statements more than one stage above their own.[4] Some individuals comprehend the next stage above their own; some do not. Adolescents prefer (or rank as best) the highest stage they can comprehend.

To understand moral stages, it is important to clarify their relations to stage of logic or intelligence, on the one hand, and to moral behavior on the other. Maturity of moral judgment is not highly correlated with IQ or verbal intelligence (correlations are only in the 30s, accounting for 10% of the variance). Cognitive development, in the stage sense, however, is more important for moral development than such correlations suggest. Piaget has found that after the child learns to speak there are three major stages of reasoning: the intuitive, the concrete operational, and the formal operational. At around age 7, the child enters the stage of concrete logical thought: He can make logical inferences, classify, and handle quantitative relations about concrete things. In adolescence individuals usually enter the stage of formal operations. At this stage they can reason abstractly, i.e., consider all possibilities, form hypotheses, deduce implications from hypotheses, and test them against reality.[c]

Since moral reasoning clearly is reasoning, advanced moral reasoning depends upon advanced logical reasoning; a person's logical stage puts a certain ceiling on the moral stage he can attain. A person whose logical stage is only concrete operational is limited to the preconventional moral stages (Stages 1 and 2). A person whose logical stage is only partially formal operational is limited to the conventional moral stages (Stages 3 and 4). While logical development is necessary for moral development and sets limits to it, most individuals are higher in logical stage than they are in moral stage. As an example, over 50% of late adolescents and adults are capable of full formal reasoning, but only 10% of these adults (all formal operational) display principled (Stages 5 and 6) moral reasoning.

The moral stages are *structures of moral judgment* or *moral reasoning*. *Structures* of moral judgment must be distinguished from the *content* of moral judgment. As an example, we cite responses to a dilemma used in our various studies to identify moral stage. The dilemma raises the issue of stealing a drug to save a dying woman. The inventor of the drug is selling it for 10 times what it costs him to make it. The woman's husband cannot raise the money, and the seller refuses to lower the price or wait for payment. What should the husband do?

The choice endorsed by a subject (steal, don't steal) is called the *content* of his moral judgment in the situation. His reasoning about the choice defines the structure of his moral judgment. This reasoning centers on the following 10 universal moral values or issues of concern to persons in these moral dilemmas:

1. Punishment
2. Property
3. Roles and concerns of affection
4. Roles and concerns of authority
5. Law
6. Life
7. Liberty
8. Distributive justice
9. Truth
10. Sex

A moral choice involves choosing between two (or more) of these values as they *conflict* in concrete situations of choice.

The stage or structure of a person's moral judgment defines: 1) *what* he finds valuable in each of these moral issues (life, law), i.e., how he defines the value, and 2) *why* he finds it valuable, i.e., the reasons he gives for valuing it. As an example, at Stage 1 life is valued in terms of the power or possessions of the person involved; at Stage 2, for its usefulness in satisfying the needs of the individual in question or others; at Stage 3, in terms of the individual's relations with others and their valuation of him; at Stage 4, in terms of social or religious law. Only at Stages 5 and 6 is each life seen as inherently worthwhile, aside from other considerations.

MORAL JUDGMENT VS. MORAL ACTION

Having clarified the nature of stages of moral *judgment*, we must consider the relation of moral judgment to moral *action*. If logical reasoning is a necessary but not sufficient condition for mature moral judgment, mature moral judgment is a necessary but not sufficient condition for mature moral action. One cannot follow moral principles if one does not understand (or believe in) moral principles. However, one can reason in terms of principles and not live up to these principles. As an example, Richard Krebs and I found that only 15% of students showing some principled thinking cheated as compared to 55% of conventional subjects and 70% of preconventional subjects.[5] Nevertheless, 15% of the principled subjects did cheat, suggesting that factors additional to moral judgment are necessary for principled moral reasoning to be translated into "moral action." Partly, these factors include the situation and its pressures. Partly, what happens depends upon the individual's motives and emotions. Partly, what the individual does depends upon a general sense of will, purpose, or "ego strength." As an example of the role of will or ego strength in moral behavior, we may cite the study by Krebs: Slightly more than half of his conventional subjects cheated. These subjects were also divided by a measure of attention/will. Only 26% of the "strong-willed" conventional subjects cheated; however, 74% of the "weak-willed" subjects cheated.

If maturity of moral reasoning is only one factor in moral behavior, why does the cognitive-developmental approach to moral education focus so heavily upon moral reasoning? For the following reasons:

1. Moral judgment, while only one factor in moral behavior, is the single most important or influential factor yet discovered in moral behavior.
2. While other factors influence moral behavior, moral judgment is the only distinctively *moral* factor in moral behavior. To illustrate, we noted that the Krebs study indicated that "strong-willed" conventional stage subjects resisted cheating more than "weak-willed" subjects. For those at a preconventional level of moral reasoning, however, "will" had an opposite effect. "Strong-willed" Stages 1 and 2 subjects cheated more, not less, than "weak-willed" subjects, i.e., they had the "courage of their (amoral) convictions" that it was worthwhile to cheat. "Will," then, is an important factor in moral behavior, but it is not distinctively moral; it becomes moral only when informed by mature moral judgment.
3. Moral judgment change is long-range or irreversible; a higher stage is never lost. Moral behavior as such is largely situational and reversible or "loseable" in new situations.

II. AIMS OF MORAL AND CIVIC EDUCATION

Moral psychology describes what moral development is, as studied empirically. Moral education must also consider moral philosophy, which strives to tell us what moral development ideally *ought to be*. Psychology finds an invariant sequence of moral stages; moral philosophy must be invoked to answer whether a later stage is a better stage. The "stage" of senescence and death follows the "stage" of adulthood, but that does not mean that senescence and death are better. Our claim that the latest or principled stages of moral reasoning are morally better stages, then, must rest on considerations of moral philosophy.

The tradition of moral philosophy to which we appeal is the liberal or rational tradition, in particular the "formalistic" or "deontological" tradition running from Immanuel Kant to John Rawls.[6] Central to this tradition is the claim that an adequate morality is *principled*, i.e., that it makes judgments in terms of *universal* principles applicable to all mankind. *Principles* are to be distinguished from *rules*. Conventional morality is grounded on rules, primarily "thou shalt nots" such as are represented by the Ten Commandments, prescriptions of kinds of actions. Principles are, rather, universal guides to making a moral decision. An example is Kant's "categorical imperative," formulated in two ways. The first is the maxim of respect for human personality, "Act always toward the other as an end, not as a means." The second is the maxim of universalization, "Choose only as you would be willing to have everyone choose in your situation." Principles like that of Kant's state the formal conditions of a moral choice or action. In the dilemma in which a woman is dying because a druggist refuses to release his drug for less than the stated price, the druggist is not acting morally, though he is not violating the ordinary moral rules (he is not actually stealing or murdering). But he is violating principles: He is treating the woman simply as a means to his ends of profit, and he is not choosing as he would wish anyone to choose (if the druggist were in the dying woman's place, he would not want a druggist to choose as he is choosing). Under most circumstances, choice in terms of conventional moral rules and choice in terms of principles coincide. Ordinarily, principles dictate not stealing (avoiding stealing is implied by acting in terms of a regard for others as ends and in terms of what one would want everyone to do). In a situation where stealing is the only means to save a life, however, principles contradict the ordinary rules and would dictate stealing. Unlike rules which are supported by social authority, principles are freely chosen by the individual because of their intrinsic moral validity.[d]

The conception that a moral choice is a choice made in terms of moral principles is related to the claim of liberal moral philosophy that moral principles are ultimately principles of justice. In essence, moral conflicts are conflicts between the claims of persons, and principles for resolving these claims are principles of justice, "for giving each his due." Central to justice are the demands of *liberty, equality*, and *reciprocity*. At every moral stage, there is a concern for justice. The most damning statement a school child can make about a teacher is that "he's not fair." At each higher stage, however, the conception of justice is reorganized. At Stage 1, justice is punishing the bad in terms of "an eye for an eye and a tooth for a tooth." At Stage 2, it is exchanging favors and goods in an equal manner. At Stages 3 and 4, it is treating people as they desire in terms of the conventional rules. At Stage 5, it is recognized that all rules and laws flow from justice, from a social contract between the governors and the governed designed to protect the equal rights of all. At Stage 6, personally chosen moral principles are also principles of justice, the principles any member of a society would choose for that society if he did not know what his position was to be in the society and in which he might be the least advantaged.[7] Principles chosen from this point of view are, first, the maximum liberty compatible with the like liberty of others and, second, no inequalities of goods and respect which are not to the benefit of all, including the least advantaged.

As an example of stage progression in the orientation to justice, we may take judgments about capital punishment.[8] Capital punishment is only firmly rejected at the two principled stages, when the notion of justice as vengeance or retribution is abandoned. At the sixth stage, capital punishment is not condoned even if it may have some useful deterrent effect in promoting law and order. This is because it is not a punishment we would choose for a society if we assumed we had as much chance of being born into the position of a criminal or murderer as being born into the position of a law abider.

Why are decisions based on universal principles of justice better decisions? Because they are decisions on which all moral men could agree. When decisions are based on conventional moral rules, men will disagree, since they adhere to conflicting systems of rules dependent on culture and social position. Throughout history men have killed one another in the name of conflicting moral rules and values, most recently in Vietnam and the Middle East. Truly moral or just resolutions of conflicts require principles which are, or can be, universalizable.

ALTERNATIVE APPROACHES

We have given a philosophic rationale for stage advance as the aim of moral education. Given this rationale, the developmental approach to moral education can avoid the problems inherent in the other two major approaches to moral education. The first alternative approach is that of indoctrinative moral education, the preaching and imposition of the rules and values of the teacher and his culture on the child. In America, when this indoctrinative approach has been developed in a systematic manner, it has usually been termed "character education."

Moral values, in the character education approach, are preached or taught in terms of what may be called the "bag of virtues." In the classic studies of character by Hugh Hartshorne and Mark May, the virtues chosen were honesty, service, and self-control.[9] It is easy to get superficial consensus on such a bag of virtues—until one examines in detail the list of virtues involved and the details of their definition. Is the Hartshorne and May bag more adequate than the Boy Scout bag (a Scout should be honest, loyal, reverent, clean, brave, etc.)? When one turns to the details of defining each virtue, one finds equal uncertainty or difficulty in reaching consensus. Does honesty mean one should not steal to save a life? Does it mean that a student should not help another student with his homework?

Character education and other forms of indoctrinative moral education have aimed at teaching universal values (it is assumed that honesty or service are desirable traits for all men in all societies), but the detailed definitions used are relative; they are defined by the opinions of the teacher and the conventional culture and rest on the authority of the teacher for their justification. In this sense character education is close to the unreflective valuings by teachers which constitute the hidden curriculum of the school.[e] Because of the current unpopularity of indoctrinative approaches to moral education, a family of approaches called "values clarification" has become appealing to teachers. Values clarification takes the first step implied by a rational approach to moral education: the eliciting of the child's own judgment or opinion about issues or situations in which values conflict, rather than imposing the teacher's opinion on him. Values clarification, however, does not attempt to go further than eliciting awareness of values; it is assumed that becoming more self-aware about one's values is an end in itself. Fundamentally, the definition of the end of values education as self-awareness derives from a belief in ethical relativity held by many value-clarifiers. As stated by Peter Engel, "One must contrast value clarification and value inculcation. Value clarification implies the principle that in the consideration of values there is no single correct answer." Within these premises of "no correct answer," children are to discuss moral dilemmas in such a way as to reveal different values and discuss their value differences with each other. The teacher is to stress that "our values are different," not that one value is more adequate than others. If this program is systematically followed, students will themselves become relativists, believing there is no "right" moral answer. For instance, a student caught cheating might argue that he did nothing wrong, since his own hierarchy of values, which may be different from that of the teacher, made it right for him to cheat.

Like values clarification, the cognitive-developmental approach to moral education stresses open or Socratic peer discussion of value dilemmas. Such discussion, however, has an aim: stimulation of movement to the next stage of moral reasoning. Like values clarification, the developmental approach opposes indoctrination. Stimulation of movement to the next stage of reasoning is not indoctrinative, for the following reasons:

1. Change is in the way of reasoning rather than in the particular beliefs involved.
2. Students in a class are at different stages; the aim is to aid movement of each to the next stage, not convergence on a common pattern.
3. The teacher's own opinion is neither stressed nor invoked as authoritative. It enters in only as one of many opinions, hopefully one of those at a next higher stage.
4. The notion that some judgments are more adequate than others is communicated. Fundamentally, however, this means that the student is encouraged to articulate a position which seems most adequate to him and to judge the adequacy of the reasoning of others.

In addition to having more definite aims than values clarification, the moral development approach restricts value education to that which is moral or, more specifically, to justice. This is for two reasons. First, it is not clear that the whole realm of personal, political, and religious values is a realm which is nonrelative, i.e., in which there are universals and a direction of development. Second, it is not clear that the public school has a right or mandate to develop values in general.[f] In our view, value education in the public schools should be restricted to that which the school has the right and mandate to develop: an awareness of justice, or of the rights of others in our Constitutional system. While the Bill of Rights prohibits the teaching of religious beliefs, or of specific value systems, it does not prohibit the teaching of the awareness of rights and principles of justice fundamental to the Constitution itself.

When moral education is recognized as centered in justice and differentiated from value education or affective education, it becomes apparent that moral and civic education are much the same thing. This equation, taken for granted by the classic philosophers of education from Plato and Aristotle to Dewey, is basic to our claim that a concern for moral education is central to the educational objectives of social studies.

The term *civic education* is used to refer to social studies as more than the study of the facts and concepts of social science, history, and civics. It is education for the analytic understanding, value principles, and motivation necessary for a citizen in a democracy if democracy is to be an effective process. It is political education. Civic or political education means the stimulation of development of more advanced patterns of reasoning about political and social decisions and their implementation in action. These patterns are patterns of moral reasoning. Our studies show that reasoning and decision making about political decisions are directly derivative of broader patterns of moral reasoning and decision making. We have interviewed high school and college students about concrete political situations involving laws to govern open housing, civil disobedience for peace in

Vietnam, free press rights to publish what might disturb national order, and distribution of income through taxation. We find that reasoning on these political decisions can be classified according to moral stage and that an individual's stage on political dilemmas is at the same level as on nonpolitical moral dilemmas (euthanasia, violating authority to maintain trust in a family, stealing a drug to save one's dying wife). Turning from reasoning to action, similar findings are obtained. In 1963 a study was made of those who sat in at the University of California, Berkeley, administration building and those who did not in the Free Speech Movement crisis. Of those at Stage 6, 80% sat in, believing that principles of free speech were being compromised, and that all efforts to compromise and negotiate with the administration had failed. In contrast, only 15% of the conventional (Stage 3 or Stage 4) subjects sat in. (Stage 5 subjects were in between.)[g]

From a psychological side, then, political development is part of moral development. The same is true from the philosophic side. In the *Republic*, Plato sees political education as part of a broader education for moral justice and finds a rationale for such education in terms of universal philosophic principles rather than the demands of a particular society. More recently, Dewey claims the same.

In historical perspective, America was the first nation whose government was publicly founded on postconventional principles of justice, rather than upon the authority central to conventional moral reasoning. At the time of our founding, postconventional or principled moral and political reasoning was the possession of the minority, as it still is. Today, as in the time of our founding, the majority of our adults are at the conventional level, particularly the "law and order" (fourth) moral stage. (Every few years the Gallup Poll circulates the Bill of Rights unidentified, and every year it is turned down.) The Founding Fathers intuitively understood this without benefit of our elaborate social science research; they constructed a document designing a government which would maintain principles of justice and the rights of man even though principled men were not the men in power. The machinery included checks and balances, the independent judiciary, and freedom of the press. Most recently, this machinery found its use at Watergate. The tragedy of Richard Nixon, as Harry Truman said long ago, was that he never understood the Constitution (a Stage 5 document), but the Constitution understood Richard Nixon.[h]

Watergate, then, is not some sign of moral decay of the nation, but rather of the fact that understanding and action in support of justice principles are still the possession of a minority of our society. Insofar as there is moral decay, it represents the weakening of conventional morality in the face of social and value conflict today. This can lead the less fortunate adolescent to fixation at the preconventional level, the more fortunate to movement to principles. We find a larger proportion of youths at the principled level today than was the case in their fathers' day, but also a larger proportion at the preconventional level. Given this state, moral and civic education in the schools becomes a more urgent task. In the high school today, one often hears both preconventional adolescents and those beginning to move beyond convention sounding the same note of disaffection for the school. While our political institutions are in principle Stage 5 (i.e., vehicles for maintaining universal rights through the democratic process), our schools have traditionally been Stage 4 institutions of convention and authority. Today more than ever, democratic schools systematically engaged in civic education are required.

Our approach to moral and civic education relates the study of law and government to the actual creation of a democratic school in which moral dilemmas are discussed and resolved in a manner which will stimulate moral development.

PLANNED MORAL EDUCATION

For many years, moral development was held by psychologists to be primarily a result of family upbringing and family conditions. In particular, conditions of affection and authority in the home were believed to be critical, some balance of warmth and firmness being optimal for moral development. This view arises if morality is conceived as an internalization of the arbitrary rules of parents and culture, since such acceptance must be based on affection and respect for parents as authorities rather than on the rational nature of the rules involved.

Studies of family correlates of moral stage development do not support this internalization view of the conditions for moral development. Instead, they suggest that the conditions for moral development in homes and schools are similar and that the conditions are consistent with cognitive-developmental theory. In the cognitive-developmental view, morality is a natural product of a universal human tendency toward empathy or role taking, toward putting oneself in the shoes of other conscious beings. It is also a product of a universal human concern for justice, for reciprocity or equality in the relation of one person to another. As an example, when my son was 4, he became a morally principled vegetarian and refused to eat meat, resisting all parental persuasion to increase his protein intake. His reason was, "It's bad to kill animals." His moral commitment to vegetarianism was not taught or acquired from parental authority; it was the result of the universal tendency of the young self to project its consciousness and values into other living things, other selves. My son's vegetarianism also involved a sense of justice, revealed when I read him a book about Eskimos in which a real hunting expedition was described. His response was to say, "Daddy, there is one kind of meat I would eat—Eskimo meat. It's all right to eat

Eskimos because they eat animals." This natural sense of justice or reciprocity was Stage 1—an eye for an eye, a tooth for a tooth. My son's sense of the value of life was also Stage 1 and involved no differentiation between human personality and physical life. His morality, though Stage 1, was, however, natural and internal. Moral development past Stage 1, then, is not an internalization but the reconstruction of role taking and conceptions of justice toward greater adequacy. These reconstructions occur in order to achieve a better match between the child's own moral structures and the structures of the social and moral situations he confronts. We divide these conditions of match into two kinds: those dealing with moral discussions and communication and those dealing with the total moral environment or atmosphere in which the child lives.

In terms of moral discussion, the important conditions appear to be:

1. Exposure to the next higher stage of reasoning
2. Exposure to situations posing problems and contradictions for the child's current moral structure, leading to dissatisfaction with his current level
3. An atmosphere of interchange and dialogue combining the first two conditions, in which conflicting moral views are compared in an open manner

Studies of families in India and America suggest that morally advanced children have parents at higher stages. Parents expose children to the next higher stage, raising moral issues and engaging in open dialogue or interchange about such issues.[10]

Drawing on this notion of the discussion conditions stimulating advance, Moshe Blatt conducted classroom discussions of conflict-laden hypothetical moral dilemmas with four classes of junior high and high school students for a semester.[11] In each of these classes, students were to be found at three stages. Since the children were not all responding at the same stage, the arguments they used with each other were at different levels. In the course of these discussions among the students, the teacher first supported and clarified those arguments that were one stage above the lowest stage among the children; for example, the teacher supported Stage 3 rather than Stage 2. When it seemed that these arguments were understood by the students, the teacher then challenged that stage, using new situations, and clarified the arguments one stage above the previous one: Stage 4 rather than Stage 3. At the end of the semester, all the students were retested; they showed significant upward change when compared to the controls, and they maintained the change one year later. In the experimental classrooms, from one-fourth to one-half of the students moved up a stage, while there was essentially no change during the course of the experiment in the control group.

Given the Blatt studies showing that moral discussion could raise moral stage, we undertook the next step: to see if teachers could conduct moral discussions in the course of teaching high school social studies with the same results. This step we took in cooperation with Edwin Fenton, who introduced moral dilemmas in his ninth- and eleventh-grade social studies texts. Twenty-four teachers in the Boston and Pittsburgh areas were given some instruction in conducting moral discussions around the dilemmas in the text. About half of the teachers stimulated significant developmental change in their classrooms—upward stage movement of one-quarter to one-half a stage. In control classes using the text but no moral dilemma discussions, the same teachers failed to stimulate any moral change in the students. Moral discussion, then, can be a usable and effective part of the curriculum at any grade level. Working with filmstrip dilemmas produced in cooperation with Guidance Associates, second-grade teachers conducted moral discussions yielding a similar amount of moral stage movement.

Moral discussion and curriculum, however, constitute only one portion of the conditions stimulating moral growth. When we turn to analyzing the broader life environment, we turn to a consideration of the *moral atmosphere* of the home, the school, and the broader society. The first basic dimension of social atmosphere is the role-taking opportunities it provides, the extent to which it encourages the child to take the point of view of others. Role taking is related to the amount of social interaction and social communication in which the child engages, as well as to his sense of efficacy in influencing attitudes of others. The second dimension of social atmosphere, more strictly moral, is the level of justice of the environment or institution. The justice structure of an institution refers to the perceived rules or principles for distributing rewards, punishments, responsibilities, and privileges among institutional members. This structure may exist or be perceived at any of our moral stages. As an example, a study of a traditional prison revealed that inmates perceived it as Stage 1, regardless of their own level.[12] Obedience to arbitrary command by power figures and punishment for disobedience were seen as the governing justice norms of the prison. A behavior-modification prison using point rewards for conformity was perceived as a Stage 2 system of instrumental exchange. Inmates at Stage 3 or 4 perceived this institution as more fair than the traditional prison, but not as fair in their own terms.

These and other studies suggest that a higher level of institutional justice is a condition for individual development of a higher sense of justice. Working on these premises, Joseph Hickey, Peter Scharf, and I worked with guards and inmates in a women's prison to create a more just community.[13] A social contract was set up in which guards and inmates each had a vote of one and in which rules were made and conflicts resolved through discussions of fairness and a democratic vote in a community meeting. The program has been operating four years and has stimulated moral stage advance in inmates, though it is still too early to draw conclusions as to its overall long-range effectiveness for rehabilitation.

One year ago, Fenton, Ralph Mosher, and I received a grant from the Danforth Foundation (with additional support from the Kennedy Foundation) to make moral education a living matter in two high schools in the Boston area (Cambridge and Brookline) and two in Pittsburgh. The plan had two components. The first was training counselors and social studies and English teachers in conducting moral discussions and making moral discussion an integral part of the curriculum. The second was establishing a just community school within a public high school.

We have stated the theory of the just community high school, postulating that discussing real-life moral situations and actions as issues of fairness and as matters for democratic decision would stimulate advance in both moral reasoning and moral action. A participatory democracy provides more extensive opportunities for role taking and a higher level of perceived institutional justice than does any other social arrangement. Most alternative schools strive to establish a democratic governance, but none we have observed has achieved a vital or viable participatory democracy. Our theory suggested reasons why we might succeed where others failed. First, we felt that democracy had to be a central commitment of a school, rather than a humanitarian frill. Democracy as moral education provides that commitment. Second, democracy in alternative schools often fails because it bores the students. Students prefer to let teachers make decisions about staff, courses, and schedules, rather than to attend lengthy, complicated meetings. Our theory said that the issues a democracy should focus on are issues of morality and fairness. Real issues concerning drugs, stealing, disruptions, and grading are never boring if handled as issues of fairness. Third, our theory told us that if large democratic community meetings were preceded by small-group moral discussion, higher-stage thinking by students would win out in later decisions, avoiding the disasters of mob rule.[i]

Currently, we can report that the school based on our theory makes democracy work or function where other schools have failed. It is too early to make any claims for its effectiveness in causing moral development, however.

Our Cambridge just community school within the public high school was started after a small summer planning session of volunteer teachers, students, and parents. At the time the school opened in the fall, only a commitment to democracy and a skeleton program of English and social studies had been decided on. The school started with six teachers from the regular school and 60 students, 20 from academic professional homes and 20 from working-class homes. The other 20 were dropouts and troublemakers or petty delinquents in terms of previous record. The usual mistakes and usual chaos of a beginning alternative school ensued. Within a few weeks, however, a successful democratic community process had been established. Rules were made around pressing issues: disturbances, drugs, hooking. A student discipline committee or jury was formed. The resulting rules and enforcement have been relatively effective and reasonable. We do not see reasonable rules as ends in themselves, however, but as vehicles for moral discussion and an emerging sense of community. This sense of community and a resulting morale are perhaps the most immediate signs of success. This sense of community seems to lead to behavior change of a positive sort. An example is a 15-year-old student who started as one of the greatest combinations of humor, aggression, light-fingeredness, and hyperactivity I have ever known. From being the principal disturber of all community meetings, he has become an excellent community meeting participant and occasional chairman. He is still more ready to enforce rules for others than to observe them himself, yet his commitment to the school has led to a steady decrease in exotic behavior. In addition, he has become more involved in classes and projects and has begun to listen and ask questions in order to pursue a line of interest.

We attribute such behavior change not only to peer pressure and moral discussion but to the sense of community which has emerged from the democratic process in which angry conflicts are resolved through fairness and community decision. This sense of community is reflected in statements of the students to us that there are no cliques—that the blacks and the whites, the professors' sons and the project students, are friends. These statements are supported by observation. Such a sense of community is needed where students in a given classroom range in reading level from fifth-grade to college.

Fenton, Mosher, the Cambridge and Brookline teachers, and I are now planning a four-year curriculum in English and social studies centering on moral discussion, on role taking and communication, and on relating the government, laws, and justice system of the school to that of the American society and other world societies. This will integrate an intellectual curriculum for a higher level of understanding of society with the experiential components of school democracy and moral decision.

There is very little new in this—or in anything else we are doing. Dewey wanted democratic experimental schools for moral and intellectual development 70 years ago. Perhaps Dewey's time has come.

NOTES

1. John Dewey, "What Psychology Can Do for the Teacher," in Reginald Archambault, ed., *John Dewey on Education: Selected Writings* (New York: Random House, 1964).

2. Jean Piaget, *The Moral Judgment of the Child*, 2nd ed. (Glencoe, Ill.: Free Press, 1948).
3. Lawrence Kohlberg, "Moral Stages and Moralization: The Cognitive-Developmental Approach," in Thomas Lickona, ed., *Man, Morality, and Society* (New York: Holt, Rinehart, and Winston, in press).
4. James Rest, Elliott Turiel, and Lawrence Kohlberg, "Relations Between Level of Moral Judgment and Preference and Comprehension of the Moral Judgment of Others," *Journal of Personality*, vol. 37, 1969, pp. 225–52, and James Rest, "Comprehension, Preference, and Spontaneous Usage in Moral Judgment," in Lawrence Kohlberg, ed., *Recent Research in Moral Development* (New York: Holt, Rinehart, and Winston, in preparation).
5. Richard Krebs and Lawrence Kohlberg, "Moral Judgment and Ego Controls as Determinants of Resistance to Cheating," in Lawrence Kohlberg, ed., *Recent Research*.
6. John Rawls, *A Theory of Justice* (Cambridge, Mass.: Harvard University Press, 1971).
7. John Rawls, ibid.
8. Lawrence Kohlberg and Donald Elfenbein, "Development of Moral Reasoning and Attitudes Toward Capital Punishment," *American Journal of Orthopsychiatry*, Summer, 1975.
9. Hugh Hartshorne and Mark May, *Studies in the Nature of Character: Studies in Deceit*, vol. 1; *Studies in Service and Self-Control*, vol. 2; *Studies in Organization of Character*, vol. 3 (New York: Macmillan, 1928–30).
10. Bindu Parilch, "A Cross-Cultural Study of Parent-Child Moral Judgment," unpublished doctoral dissertation, Harvard University, 1975.
11. Moshe Blatt and Lawrence Kohlberg, "Effects of Classroom Discussions upon Children's Level of Moral Judgment," in Lawrence Kohlberg, ed., *Recent Research*.
12. Lawrence Kohlberg, Peter Scharf, and Joseph Hickey, "The Justice Structure of the Prison: A Theory and an Intervention," *The Prison Journal*, Autumn-Winter, 1972.
13. Lawrence Kohlberg, Kelsey Kauffman, Peter Scharf, and Joseph Hickey, *The Just Community Approach to Corrections: A Manual, Part I* (Cambridge, Mass.: Education Research Foundation, 1973).

a. These levels correspond roughly to our three major levels: the preconventional, the conventional, and the principled. Similar levels were propounded by William McDougall, Leonard Hobhouse, and James Mark Baldwin.

b. Piaget's stages correspond to our first three stages: Stage 0 (pre-moral), Stage 1 (heteronomous), and Stage 2 (instrumental reciprocity).

c. Many adolescents and adults only partially attain the stage of formal operations. They do consider all the actual relations of one thing to another at the same time, but they do not consider all possibilities and form abstract hypotheses. A few do not advance this far, remaining "concrete operational."

d. Not all freely chosen values or rules are principles, however. Hitler chose the "rule," "exterminate the enemies of the Aryan race," but such a rule is not a universalizable principle.

e. As an example of the "hidden curriculum," we may cite a second-grade classroom. My son came home from this classroom one day saying he did not want to be "one of the bad boys." Asked "Who are the bad boys?" he replied, "The ones who don't put their books back and get yelled at."

f. Restriction of deliberate value education to the moral may be clarified by our example of the second-grade teacher who made tidying up of books a matter of moral indoctrination. Tidiness is a value, but it is not a moral value. Cheating is a moral issue, intrinsically one of fairness. It involves issues of violation of trust and taking advantage. Failing to tidy the room may under certain conditions be an issue of fairness, when it puts an undue burden on others. If it is handled by the teacher as a matter of cooperation among the group in this sense, it is a legitimate focus of deliberate moral education. If it is not, it simply represents the arbitrary imposition of the teacher's values on the child.

g. The differential action of the principled subjects was determined by two things. First, they were more likely to judge it right to violate authority by sitting in. But second, they were also in general more consistent in engaging in political action according to their judgment. Ninety percent of all Stage 6 subjects thought it right to sit in, and all 90% lived up to this belief. Among the Stage 4 subjects, 45% thought it right to sit in, but only 33% lived up to this belief by acting.

h. No public or private word or deed of Nixon ever rose above Stage 4, the "law and order" stage. His last comments in the White House were of wonderment that the Republican Congress could turn on him after so many Stage 2 exchanges of favors in getting them elected.

i. An example of the need for small-group discussion comes from an alternative school community meeting called because a pair of the students had stolen the school's video-recorder. The resulting majority decision was that the school should buy back the recorder from the culprits through a fence. The teachers could not accept this decision and returned to a more authoritative approach. I believe if the moral reasoning of students urging this solution had been confronted by students at a higher stage, a different decision would have emerged.

19. A Critical Study of Kohlberg's Theory of the Development of Moral Judgments

Betty A. Sichel

Lawrence Kohlberg's theory of the development of moral judgments together with his proposals for moral education have found a ready audience among psychologists, educators, and even philosophers. However, this audience has not been docile, ready to dispense bouquets of favorable comments. If anything, often the bouquets have been laced with thorns. The applause has been tempered with voices critical of various aspects of the developmental theory and educational proposals. Since a flurry of criticism has not moved Kohlberg to alter his scenario, one may wonder whether additional criticism is warranted.[1] No matter the many theoretical and practical problems as yet unanswered and unresolved by Kohlberg, the importance of Kohlberg's work cannot be minimized. In the end, whether corrections or modifications of Kohlberg's theory fall on Kohlberg's shoulders or on another pair of sturdier shoulders is immaterial. What may assist such an enterprise, however, is a thorough analysis of Kohlberg's theory of the development of moral judgments.

In this paper, I will attempt to avoid those criticisms already leveled against Kohlberg's theory.[2] In a set of well developed essays, Peters has questioned aspects of Kohlberg's theory, e.g., the lack of attention to the affective domain, the function of habit in moral development, the relationship between virtues and principles.[3] Alston has questioned whether Kohlberg provides more than empirical evidence for a hierarchy of stages and whether adequate theoretical analysis exists for the assertion that lower stages are logically necessary conditions for the existence and development

Reprinted, by permission, from *Philosophy of Education 1976*, Proceedings of the Thirty-second Annual Meeting of the Philosophy of Education Society, 1976, pp. 209–220.

of higher stages.[4] Crittenden and Edel each have been concerned with Kohlberg's minimization of the importance of the content of moral judgments as this content varies in different cultures. To enunciate elements, modes, and principles of the content of moral judgments, as Kohlberg does, is a far cry from the actual content of moral life within exceptionally divergent cultures.[5] Except to negate the validity of an approach based on the extreme relativism of diverse moral content, Kohlberg ignores aspects of content which may be vital for moral development and moral education.[6] Lastly, questions may be raised concerning the theoretical grounding and analysis of concepts within each of the six stages. For example, Kohlberg is not concerned with the nature of autonomy or with the degree of autonomy which should or could exist in any human life. Baier's analysis of the concept 'autonomy' points to some problems which may confront Kohlberg.[7]

If Kohlberg had merely carved out moral development as one portion of psychology to study, he might be on fairly safe theoretical grounds.[8] However, once he trampled over the boundaries of moral psychology into the realms of moral philosophy and moral education, other theoretical dimensions required consideration. My critical analysis will focus on concepts and ideas which have not been adequately analyzed in Kohlberg's theory and which must be included for its extension into moral philosophy and moral education: (1) The relationship between moral judgments and moral behavior. (2) Differences between logical judgments and moral judgments. (3) The inception of moral reasoning. (4) Moral justification and moral judgments.

I. THE RELATIONSHIP BETWEEN MORAL JUDGMENTS AND MORAL BEHAVIOR

Though formal education should be concerned with moral judgments, it must equally be concerned with moral behavior. This is not to claim that formal education should dogmatically prescribe one particular type of moral behavior, but rather that in some manner, moral behavior must be a concern of formal education. It would be rather ludicrous if children became verbally glib Stage Six individuals without any manifestation of their verbal judgments in actual moral behavior. Kohlberg's recognition of the importance of moral behavior[9] has not caused theoretical analysis concomitant with the complexity of the problem. If his moral development theory is to be applied to moral education, as Kohlberg asserts, far more attention must be directed towards the problem of moral behavior and the relationship between moral judgments and moral behavior.

Throughout the history of human thought the relationship between moral judgments and moral behavior has remained a puzzle. In Euripides' "Hippolytus," Phaedra tells of the chasm between moral judgment and moral behavior:

> Many a time in night's long empty spaces / I have pondered on the causes of life's shipwreck. / I think that our lives are worse than the mind's quality / would warrant. There are many who know virtue. / We know the good, we apprehend it clearly. / But we can't bring it to achievement. / [10]

Concern for the apparent slippage between moral judgments and moral deeds does not uniquely bother Euripides alone. Plato and Aristotle expend considerable energy on the problem of *akrasia* or weakness of will.[11] Throughout the ages scholars have puzzled over the merits of classical resolutions of the problem of weakness of will. Moving from classical to twentieth century analyses, e.g., by Hare, Davidson, Nowell-Smith,[12] provides little consolation for anyone searching for resolution of this knotty problem. Has Kohlberg resolved a problem which has perplexed human beings, whether common man or philosopher? Actually, Kohlberg barely touches the problem of *akrasia* or weakness of will, i.e., whether moral behavior necessarily follows moral judgment. This is not to say that Kohlberg's theory could not accommodate a resolution of the problem, merely that, at present, it does not. My focus on Socrates' treatment of *akrasia* will delineate the scope of the problem and will consider the adequacy of one possible resolution of the problem for Kohlberg.

The problem of *akrasia* or weakness of will, as considered here, can be stated in the following way:

> (a) In a moral situation, an agent, according to available knowledge, judges X to be the best, most desirable course of action. However, this agent's intentional action is consistent with what he has judged to be a less desirable action, Y.[13]

Socrates' denial of *akrasia* is baffling in that he asserts that an individual would not knowingly do wrong, evil, or what is a lesser good. If an individual knows what is good, then he will act upon that knowledge and do what is good.[14] Actually, Socrates is asserting perfect consistency between word and deed, between moral judgment of the good in any concrete situation and actual moral action. *Akrasia,* in the form presented above, will not exist for Socrates in that any agent judging X to be best, will act upon X. However, how does Socrates explain the commonsensical idea that individuals do seem to do Y, even though X was judged better,[15] or even though in retrospect, Y is regretted? In *Protagoras*, Socrates recognizes this problem:

> . . . the rest of the world are of the opinion that knowledge is not a powerful, lordly, commanding thing; they do not think of it as actually being anything of that sort at all, but their notion is that a man may have knowledge, and yet that the knowledge which is in him may be overmastered by anger, or pleasure, or pain or love, or perhaps, by fear—just as if knowledge were nothing but a slave and might be dragged about by all of these other things.[16]

After rejecting this opinion of the rest of the world, that *akrasia* is caused by emotions, Socrates asserts that *akrasia* is caused by miscalculation, error in judgment, underestimation or overestimation of short term goods *vis-à-vis* long term goods.[17] If an agent judges the comparative worth of two goods, one good is deemed better than the other. After acting on this good, the individual in retrospect, realizes that the other good should have been desired and evaluated best. The problem here is one of inadequate knowledge. For example, there is the possibility of courageously moving against an enemy or abating one's fears by retreating. After choosing to abate one's fears through retreat, as the good in this situation, the agent is court-martialed and condemned to death. In retrospect, the other alternative, courageously marching into battle, seems the actual good for that situation, especially since the battle was won and all of the participants received medals for bravery. It was not the emotion 'fear' that caused *akrasia*, but rather the *judgment* that the abatement of fear was the better good. Thus, Socrates' resolution of the problem of *akrasia* is not dependent on the emotional, but on the cognitive. Even a superficial glance reveals that Socrates invests knowledge with incredible power. Knowledge, for Socrates,

> . . . is a noble thing and fit to command in man, which cannot be overcome and will not allow a man, if only he knows the good and the evil, to do anything which is contrary to what his knowledge bids him to do.[18]

If an individual's knowledge of the good and just is absolute, most certainly the individual will not scorn this for something judged evil, unjust, or undesirable.[19] However, even without absolute knowledge, an individual will use the knowledge at his disposal in a particular concrete situation to judge what is better, more desirable, and even more pleasurable. Since this knowledge is not absolute, errors can occur in calculation, in determination of long term and short term goods, or in the underestimation or overestimation of the worth of any particular good. Thus, for Socrates, *akrasia*, as posited in (a) does not occur. Instead:

> (b) In a moral situation, an agent, according to available knowledge, judges X to be the best, most desirable course of action. This agent's intentional action will be consistent with X. If the agent's intentional action is consistent with Y, then

prior to action, reconsideration of the problem, in the light of further knowledge, caused *Y* to be judged better than *X*.

Kohlberg's theory can claim to resolve the problem of *akrasia* in a manner analogous to the Socratic resolution. Kohlberg assumes consistency between moral behavior and moral judgment, with moral judgment characteristic of the individual's stage of moral development. Virtue is knowledge, for Kohlberg; however, knowledge is characteristic of the individual's stage of moral development. For example, if Heinz,[20] a stage six individual, makes the judgment to steal the drug for his wife because of the sanctity of human life, then Heinz would be expected to follow this judgment with consistent moral behavior. However, Kohlberg does accept the idea that slippage from the highest stage of an individual's moral development to a lower stage is possible. If this were to occur, of course, action would also follow the lower stage judgment. For example, in a similar case, a close friend pleads with Heinz, our tested stage six individual, to steal a drug to save the friend's life.[21] In the battle of goods and rights, Heinz judges that the risk is too great; he has already escaped once, when stealing the drug for his wife; he is concerned with the consequences; concerned with what neighbors and family will think of him; and finally, he remembers that his wife did not show proper gratitude for his theft; she died, after changing her will in favor of her lover. Since the good in this case does not reflect stage six, but a lower stage, perhaps, stage three, there is slippage from the stage at which Heinz could operate. In this case, (b) would have to be restated in the following manner:

> (c) Let's suppose, an agent is expected to judge *X* the best, most desirable course of action, both according to available knowledge and consistent with the particular stage he is at. This agent's intentional action should also be consistent with *X*. If, however, the agent considers *Y* better, and the judgment is consistent with a *lower* stage of moral judgment, slippage has occurred. In the conflict between competing goods, *X* and *Y*, the good of the lower stage of moral development was chosen.

In retrospect, e.g., after someone else stole the drug to save the friend, the agent may question his own judgment. He may realize that *X* was actually the better judgment for reasons consistent with stage six principles. However, at the moment of the actual judgment, given the fact that he could reason at stage six, miscalculation, underestimation or overestimation caused it to fall to a stage three judgment.

Even if this presentation were expanded, there still would be compelling reasons why Socrates' resolution of *akrasia* may not be one which is accessible to Kohlberg. First, when Kohlberg allows for the possibility of slippage and thus, opens the door to a possible resolution of *akrasia*, he undermines a fundamental aspect of his theory. It is expected that once a person throws off a lower stage and moves into a higher stage, his moral judgments will reflect the higher stage. In other words, it seems inimicable with a developmental theory, such as Kohlberg's, for an individual to will a lower stage of judgments or to revert to lower stages. If Kohlberg is willing to allow for this possibility, he no longer posits hierarchical structures of thought through which an individual passes, but merely conflicting principles and conflicting moral goods. These conflicting goods may, in his opinion, be a hierarchy of comparative goods, but if contextually changeable, they in no way indicate a developmental theory.[22] For Socrates, the individual with absolute knowledge will *always* manifest knowledge of the good and just in moral action. For Kohlberg, the stage six individual, in possession of the highest principles, can still experience slippage to a lower stage.

Second, when Socrates asserts that miscalculation, underestimation, or overestimation of goods, is the cause of what often seems to be *akrasia*, the term 'good(s)' is used with exceptionally wide conceptual meaning. "Good(s)' refers to everything desired and sought by man, everything which will afford pleasure and happiness, the useful and the good life. Even if extreme abnormalities of taste, e.g., the goods sought by a Marquis de Sade, were eliminated, Socrates' use of the term 'good(s)' extends well beyond Kohlberg's use of similar moral concepts, e.g., justice.[23] Though Kohlberg's descriptive and prescriptive moral development theory[24] indicates the nature of moral life, he gives no suggestion as to the ultimately good life. In contradistinction to this, all of the lower goods, for Socrates, are finally referrable to an ultimate good which will guarantee the best, happiest, and most pleasurable of lives. Is stage six the happiest and most pleasurable of all possible lives? Without question, it is possible to posit other goods or other values as characteristic of the good and the happy life.[25] Until Kohlberg can demonstrate why a stage six life is the ultimate good life, the happiest, and most pleasurable of lives, he has not considered all aspects of human life.

Third, when Socrates analyzes *akrasia*, he is referring to the problems posed by (a) and resolved through (b). In other words, Socrates concern is not merely with moral judgments, but with the relationship between moral judgments and moral action. However, by accepting the possibility of (c), Kohlberg stresses volition in relation to judgment and relegates moral action to a secondary role. His primary concern is to note that there was slippage from a higher stage to a lower stage. Thus, the thrust of the two concerns, that of Socrates and that of Kohlberg, seem quite different.

Devoid of the conceptual framework necessary to follow Socrates' resolution of the problem of *akrasia*, Kohlberg cannot resort to this solution. Without explicit modification of the Socratic theory or some other solution, Kohlberg has avoided the problem of the relationship between moral judgment and moral behavior. However, avoidance of this problem may strand the theory in the realm of the purely theoretical, without any practical implications for actual human life. Since Kohlberg is advocating a theory which is to have implications for formal moral education, this state of affairs would be intolerable.

II. DIFFERENCES BETWEEN LOGICAL JUDGMENTS AND MORAL JUDGMENTS

Kohlberg's acceptance of Piaget's cognitive developmental theory as the paradigm for his moral developmental theory is apparent. However, application of the Piagetean form of cognitive judgment to moral judgments creates a problem. Whereas, often a logical or mathematical judgment in itself provides a necessary condition for equivalent logical behavior, a moral judgment in itself does not provide a necessary condition for equivalent moral behavior.[26] This point can be illustrated by means of a brief comparison between the nature of logical or mathematical cognitive judgments and the nature of moral cognitive judgments.

If a mathematical problem is presented, an individual may be asked *how* to solve the problem. For example, with a second degree equation, without performing any actual operations, the response could be: "By factoring." When asked "Why?", the individual may cite all sorts of theoretical principles concerning the nature of quadratic equations and factoring. Now, the individual may be prodded further: "Are you really certain that you can solve *that* problem by factoring?" "What is the correct solution of the problem?" Though technically, the "how" or "why" may involve the responses noted, the more common, expected response would be the one given to the questions just posed: "What is the actual solution of the problem?" Once the individual commits himself to a particular mathematical or logical judgment, as the correct one, the operative action or mental processes to complete the solution of the problem is necessarily implied. Thus, in the case of the second degree equation, if the roots are judged to be positive three and positive four, knowledge of this is dependent on the actual solution of the equation. At this point, since the problem is completely solved, no other action is necessary.

If a moral problem is presented, e.g., the Heinz dilemma, an individual may be asked: "Should Heinz have done that?" No matter the response, the next question will be: "Why?" At this juncture, the first response will be justified through the use of various concepts, principles, and standards. Up to this point, there is striking similarity between the moral judgment and the mathematical judgment in that mental processes exist in the making of either judgment. However, with an actual moral dilemma, rather than the hypothetical Heinz dilemma, there is a crucial difference between the mathematical or logical judgment and the moral judgment. When faced with an actual moral problem, the individual must follow his moral judgment with an additional type of behavior, actually carrying out the judgment in action. In contradistinction to this, the manifestation of the judgment in action, in the case of the mathematical or logical problem, involves the embeddedness of all operative action within the judgment itself. Since the moral behavior is not embedded within or prior to the moral judgment, this moral behavior may or may not follow the moral judgment.

Further, loosely following Aristotle's division, this distinction can be characterized as the difference between theoretical wisdom, e.g., logical or mathematical judgments, and practical wisdom with one class being moral cognitive judgments.[27] The process of evolving a practical judgment includes desire, deliberation, perception, choice, i.e., judgment, and then, finally, action. Since Aristotle's concern here is with a process to determine means, it is more likely—and yet not a necessary or sufficient condition—for an individual to manifest the means in action. Since Kohlberg's process is primarily a process of conflicting ends, the individual does not even necessarily consider means. However, whether in the case of Aristotle's practical wisdom or Kohlberg's moral cognitive judgments, moral action is a stage after choice or judgment. With logical judgments the last step is the actual choice or the judgment dependent on solution. There is no further step.

This comparison of the relationship of logical judgments to logical operations with the relationship of moral judgments to moral behavior rests on a fairly clear theoretical problem. Since a wide hiatus does not exist between logical judgments and logical operations undue energy need not be expended on bridging the hiatus. On the other hand, the gulf between moral judgments and moral behavior may be so incredibly wide that considerable theoretical energy must be expended in the hope that perhaps the gulf may be bridged. In his enthusiasm to embrace the Piagetean paradigm of cognitive development, Kohlberg has not noticed this difference between the logical and the moral. And yet this difference should raise a theorist's suspicions that the relationship between moral judgments and moral behavior is a far from simple one.

III. THE INCEPTION OF MORAL THINKING

Prior to moral reasoning how does an individual realize that he is confronted with a moral conflict? Most certainly, it is not by virtue of automatic presentation of ready-made dilemmas as occurs in empirical studies of the type conducted by Kohlberg. In a manner consistent with reflective thinking in general, moral reasoning, for Dewey, often arises out of a feeling of confusion, disharmony, or conflict, e.g., conflicting rights, goods. The emotional or physical component engendering disharmony is not purely non-cognitive in that a cognitive element determines the quality or character of the emotional or physical component. In other words, Dewey would claim that the emotional or physical is channeled, transformed, and determined by previous experiences, previous reflective thinking or moral reasoning. However, in a primary sense, even the possiblity of the realization of the existence of a moral conflict will be a function of the education of the affective by experiences and reflective thinking.

The importance of the recognition of a moral conflict cannot be underestimated. If every experience in the continuum of experience or every situation were to engender moral conflict for the indivual, some sort of neuroticism or paranoia would ensue. At the same time, the recognition of too few situations as moral conflicts would be equally unfortunate, a form of moral insensibility or moral myopia. How does an individual recognize a situation as a moral conflict or as a legitimate, important moral conflict? Though judgment as to the exact nature, components, or characteristics of the moral conflict, i.e., the intellectualization of the moral conflict, may be primarily cognitive, the affective is logically primary in motivating, causing, or generating the inception of this moral cognitive judgmental process.

Though Kohlberg views the moral problem as a conflict, he does not analyze how an individual recognizes or realizes the existence of a moral conflict. From the welter of experiences, events, and situations, the individual must sense or feel the existence of a potential moral problem. If Kohlberg believes the recognition of a moral dilemma to be a cognitive matter, evidence or analysis to sustain this is missing. If it is based on the affective, then Kohlberg must give far more attention and weight to the affective.[28] Throughout history philosophers have taken account of the affective, e.g., as a means of generating and motivating the cognitive.[29] Of course, this does not negate the cognitive influence on the affective. But it does emphasize the importance, if not the logical primacy, of the role of the affective in the generation of the cognitive. Kohlberg has not realized the subtle role of the affective for the functioning of the cognitive. In this light he has not discerned the difference between emotions which produce action and emotions which overcome the individual.[30] The former emotions may be the ones which generate the process of moral cognition and, therefore, are a necessary component of moral development theory. Other conceptual problems involving the affective are left unresolved by Kohlberg, e.g., the difference between and the relation between desires and emotions.[31]

Since situations are not presented with labels to human beings in the normal course of life, in some way, the human being must realize the existence of a moral conflict in a situation. Kohlberg has not paid attention to the means by which human beings come to the realization of a moral conflict. Formal education cannot avoid this problem.

IV. MORAL JUSTIFICATION *VIS-À-VIS* MORAL JUDGMENTS

For Kohlberg principles and standards are to serve as the means by which to determine the moral developmental stage of an individual. As is necessary for any moral principles, Kohlberg does expend some theoretical energy to clarify and analyze these principles and related concepts. When these concepts, principles, or standards are used to determine the stage of moral development for any moral judgment do the concepts, principles, and standards provide justification for the judgment after the judgment had been made? Or did the concepts, principles, and standards enter into the process of making the judgment? At times, Kohlberg seems to confuse the process of making a moral judgment with justification of a moral judgment. Often responses from Kohlberg's sample seem to be closer to justification and yet most often Kohlberg himself refers to the process of judging. The difference here is between the justification for an inference and the inferring itself or between justification of a judgment and the process of the judgment.

Individual *A*, involved in the process of resolving a moral conflict, uses various moral standards and principles as well as various modes, elements, and principles of content. In other words, first, the individual faces a moral conflict. Then, in conjunction with the content of the moral situation, principles and standards are used in the evolution of a moral judgment. However, another individual, *B*, may verbalize a position regarding a moral problem without any thought process, maybe, merely repeating what he had previously heard. A bystander may ask *B* to justify what appears to be a moral judgment. Standards, principles, and concepts will then be used and given in the attempt to justify the stated moral proposition. In this second case, *B* did not originally use standards and principles in the process of evolving the judgment, but merely in justifying his statement. In the first case, on the other hand, *A* has not only used standards to justify his judgments, but has also used them in the formation of the judgment. However, even in the case of *A* there could be some hiatus between the principles used in the formation of the judgment and those presented during the justification. Though it is hoped that the principles and standards used in the process of developing the moral judgment will be identical with the principles and standards given to justify the moral judgment, there is no necessity that this be the case. Since Kohlberg does not differentiate, in any way, between moral judgment and moral justification, one must assume that he does not recognize the subtle difference.

My emphasis on the difference between the justification of a judgment and the making of a judgment is illustrative of what I consider to be a most difficult and questionable aspect of Kohlberg's work. In the course of a series of papers written over a number of years, there is a certain amount of bandaging of Kohlberg's theory with very little detailed analytical attention to the concepts, ideas, and issues which would be necessary to qualify the theory as a candidate for educational proposals.[32] Too often ideas are merely mentioned and theoreticians are succinctly noted. However, without the rigorous, detailed analysis necessary for a full-blown moral philosophy or for formal education proposals, the theory cannot sustain even spasmodic forays into these areas. If Kohlberg's theory and educational proposals are to be more than a pleasing response to a critical, practical need, Kohlberg must begin the hard work of theoretical analysis. Avoidance of response to critics and avoidance of theoretical analysis can only involve Kohlberg in the very dogmatism and authoritarianism which he so vigorously rejects. Given the need for moral educational proposals based on valid theoretical foundations, one can only hope that Kohlberg will confront the issues which he has avoided to date.

NOTES

1. The frustration and annoyance by critics is apparent in the title of Peters' most terse criticism of Kohlberg: R. S. Peters, "A Reply to Kohlberg: Why doesn't Lawrence Kohlberg do his homework?" *Phi Delta Kappan*, Vol. LVI, no. 10 (June, 1975), p. 678.
2. In addition to those mentioned subsequently see: C. Bereiter, "Educational Implications of Kohlberg's Cognitive Developmental View," *Interchange*, Vol. 1, no. 2 (1970), pp. 25–32; John B. Orr, "Cognitive-Developmental Approaches to Moral Education: A Social Ethical Analysis," *Educational Theory*, Vol. 24, no. 4 (Fall, 1974), pp. 365–373; Kathryn Pauly Morgan, "Philosophical Problems in Cognitive-Developmental Theory: A Critique of the Work of Lawrence Kohlberg," *Proceedings of the Twenty-Ninth Annual Meeting of the Philosophy of Education Society* (ed.) B. Crittenden (Edwardsville, Ill.: Southern Illinois Press, 1973), pp. 104–117; Robert W. O'Connor and Victor L.

Worsfold, "Kohlberg's Developmental Stages as Ethical Theories: Some Doubts," *Ibid.*, pp. 118–125; for abstracts of two criticisms presented in response to Kohlberg's paper at the Seventieth Annual Meeting of the American Philosophical Association, Eastern Div.: Kurt Baier, "Individual Moral Development and Social Moral Advance," *The Journal of Philosophy*, Vol. LXX, no. 18 (1973), p. 646; Richard G. Henson, "Correlativity and Reversability," *Ibid.*, pp. 648–649; Marvin Bressler, "Kohlberg and the Resolution of Moral Conflict," *New York University Education Quarterly*, Vol. VII, no. 2 (1976), pp. 2–8.

3. R. S. Peters, "Education and Human Development," *Education and the Development of Reason* (eds.) R. F. Dearden, P. H. Hirst, and R. S. Peters (London: Routledge & Kegan Paul, 1972), pp. 501–520. R. S. Peters, "Concrete Principles and the Rational Passions," *Moral Education: Five Lectures* (Cambridge: Harvard University Press, 1970), pp. 33–37. R. S. Peters, "Moral Development: A Plea for Pluralism," *Cognitive Development and Epistemology* (ed.) T. Mischel (New York: Academic Press, 1971), pp. 237–267.
4. William P. Alston, "Comments on Kohlberg's 'From Is to Ought'," *Cognitive Development and Epistemology* (ed.) T. Mischel (New York: Academic Press, 1971), pp. 269–284.
5. Brian Crittenden, *Form and Content in Moral Education* (Toronto: The Institute for Studies in Education, 1972), pp. 14–23. Summary of A. Edel's criticism in: *Perspectives on Research: Conference on Studies of the Acquisition and Development of Values* (Bethesda, Md.: National Institute of Child Health and Human Development, 1968), p. 53.
6. Lawrence Kohlberg, "From Is to Ought: How to Commit the Naturalistic Fallacy and Get Away with It in the Study of Moral Development," *Cognitive Development and Epistemology* (ed.) T. Mischel (New York: Academic Press, 1971), "II. Universals and Relativity in Moral Development," pp. 155–180. Lawrence Kohlberg, "Indoctrination versus Relativity in Value Education," *Zygon*, Vol. 6, no. 4 (1971), pp. 285–310.
7. Kurt Baier, "Moral Autonomy as an Aim of Moral Education," *New Essays in Philosophy of Education* (eds.) G. Langford and D. J. O'Connor (London: Routledge & Kegan Paul, 1973), pp. 96–114.
8. Whether any theory with insufficient conceptual development is on safe grounds has been discussed in the following: D. W. Hamlyn, "Epistemology and Conceptual Development," *Cognitive Development and Epistemology* (ed.) T. Mischel (New York: Academic Press, 1971), pp. 3–24. Stephen Toulmin, "The Concept of 'Stages' in Psychological Development," *Ibid.*, pp. 25–60.
9. Lawrence Kohlberg, "The Cognitive-Development Approach to Moral Education," *Phi Delta Kappan*, Vol. LVI, no. 10 (1975), pp. 672–673.

10. Euripides, "Hippolytus," 375ff (David Greene, trans.). And again this theme is reiterated when Hippolytus contrasts the quality of his moral judgments and moral deeds with those of Phaedra: "Virtuous she was in deed, although not virtuous: / I that have virtue used it to my ruin. / Phaedra behaved with self control although / she had not the power to be chaste, while I who have the power, have not used it well." *Ibid.*, 1034ff.
11. For analysis of this: James J. Walsh, *Aristotle's Conception of Moral Weakness* (New York: Columbia University Press, 1963), *passim.* Gregory Vlastos, "Introduction," *Plato's "Protagoras"* (Indianapolis: Bobbs-Merrill Co., Inc., 1956), pp. vii–lvi.
12. R. M. Hare, "Freedom of the Will," *Essays on the Moral Concepts* (London: MacMillan Press, Ltd., 1972), pp. 1–12. R. M. Hare, *Freedom and Reason* (London: Oxford University Press, 1963), ch. 4, "'Ought' and 'Can'." Donald Davidson, "How is Weakness of the Will Possible?" *Moral Concepts* (ed.) Joel Feinberg (London: Oxford University Press, 1969), pp. 93–113. P. H. Nowell-Smith, *Ethics* (Baltimore: Penguin Books, 1954), pp. 265–269.
13. In conjunction with Kohlberg's theory, this statement of weakness of will is limited to moral dilemmas.
14. It is on the basis of this that Kohlberg likens his theory to that of Socrates.
15. A contemporary example of this is found in Thomas Mann's *Buddenbrooks* (H. T. Lowe-Porter, trans.) (New York: Alfred A. Knopf, 1946), pp. 417–419.
16. *Prot.*, 352B–C (Jowett-Ostwald, trans.).
17. Gregory Vlastos, "Socrates on Acrasia," *Phoenix*, Vol. 23 (1969), pp. 71–88.
18. *Prot.* 352C.
19. In his analysis of Socrates' resolution of *akrasia*, Vlastos uses Jeffrey's theory of preference and theory of decision as an example of how individuals determine comparative goods, e.g., Richard C. Jeffrey, *The Logic of Decision* (New York: McGraw-Hill, 1965).
20. The Heinz dilemma is one of the moral conflicts presented by Kohlberg to determine an indivual's moral development stage: "In Europe, a woman was near death from a very bad disease, a special kind of cancer. There was one drug that the doctors thought might save her. It was a form of radium for which a druggist was charging ten times what the drug cost him to make. The sick woman's husband, Heinz, went to everyone he knew to borrow the money, but he could only get together about half of what it cost. He told the druggist that his wife was dying, and asked him to sell it cheaper or let him pay later. But the druggist said, 'No, I discovered the drug and I'm going to make money from it.'

So Heinz got desperate and broke into the man's store to steal the drug for his wife. Should the husband have done that? Why?" Kohlberg, "From Is to Ought . . . ," p. 156.

21. Actually, the stages cannot be determined merely on the basis of whether Heinz should steal the drug for his wife. Rather, the subjects' responses to questions must involve an extension of classes of individuals considered.
22. Though it be expected that once development has occurred, the majority of judgments and/or actions will henceforth be characteristic of that stage, this does not seem to be the case: Lawrence Kohlberg, Peter Scharf, and Joseph Hickey, "The Justice Structure of the Prison—A Theory and an Intervention," *The Prison Journal*, Vol. LI (1972), pp. 3–14.
23. Regarding the elimination of abnormal goods from such a theory: Vlastos, "Socrates on . . . ," p. 74.
24. Though Kohlberg might deny that his theory is prescriptive, I believe that this assertion can be sustained.
25. Other developmental theories seem to consider goods outside of the moral sphere, e.g., those by Maslow, Erikson.
26. Though the following point is not made in this section of the paper, it is believed that it can be sustained: The form of a logical operation will not vary greatly even though it may vary. However, the form or style of moral behavior, subsequent to the judgment, may vary incredibly. In fact, the variation of style or form may be so great that it may further accent the slippage between moral judgment and moral behavior.
27. Aristotle, *N.E.*, Book VI.
28. Kohlberg's brief comments on the affective do not take this use of the affective into consideration: Kohlberg, "From Is to Ought . . . ," pp. 188–190.
29. For example, Plato uses the affective in this manner, e.g., *Rep.* V, 475A–E, VI 485C–E; *Phaedrus*, 249C ff; *Symp.* 204B ff.
30. Bernard Williams, "Morality and the Emotions," *Morality and Moral Reasoning* (ed.) J. Casey (London: Methuen & Co., Ltd., 1971), p. 19.
31. Anthony Kenny, *Action, Emotion and Will* (New York: Humanities Press, Inc., 1963), Ch. Five, "Emotion."
32. In relation to this criticism as it refers to psychology in general, Toulmin states that " . . . theoretical obscurities afflicting contemporary psychology may be seen in a clear light if looked at from the standpoint of analytical philosophy." Stephen Toulmin, "Concepts and the Explanation of Human Behavior," *Human Action: Conceptual and Empirical Issues* (ed.) T. Mischel (New York: Academic Press, 1969), p. 71. R. Harré notes that he only perceives a hint of the "origin and development

of . . . cognitive structures" in Kohlberg. R. Harré, "Some Remarks on 'Rule' as a Scientific Concept," *Understanding Other Persons* (ed.) T. Mischel (Totowa, N.J.: Rowman and Littlefield, 1974), p. 157. These comments could also be extended to cover the remainder of Kohlberg's theory and ideas.

Contributors

John L. Childs was Professor of Philosophy of Education at Teachers College, Columbia University until his retirement in 1954. He was one of the leading spokesmen for the social reconstructionist point of view in American educational philosophy. His publications include *Education and the Philosophy of Experimentalism*, *Education and Morals*, and *American Pragmatism and Education*.

John Dewey was a leading American philosopher and possibly the country's foremost educational theorist. His major works in education include *The School and Society*, *The Child and the Curriculum*, *Moral Principles in Education*, *How We Think*, *Democracy and Education*, and *Experience and Education*.

E. R. Emmet served as assistant schoolmaster for some thirty years and also as housemaster for sixteen years at Winchester College in Winchester, England. Among his books are *The Use of Reason*, *Learning to Philosophize*, *Learning to Think*, and *Handbook of Logic*.

Charles Frankel was Old Dominion Professor of Philosophy and Public Affairs at Columbia University. He also served as Assistant Secretary of State for Educational and Cultural Affairs and as the first director of the National Humanities Center. His books include *The Faith of Reason*, *The Case for Modern Man*, *The Golden Age of American Philosophy*, *The Democratic Prospect*, and *Education and the Barricades*.

William K. Frankena is Professor Emeritus of Philosophy at the University of Michigan and a leading contemporary moral philosopher. His publications include *Ethics*, *Philosophy of Education*, *Three Historical Philosophies of Education*, *Some Beliefs About Justice*, and *Thinking About Morality*. He is a former president of the American Philosophical Association.

Merrill Harmin is Professor of Education at Southern Illinois University. He is the author of *What I've Learned About Values Clarification* and the co-author of several books, including *Values and Teaching*, *The Peaceable Teacher*, and *Teaching Is. . . .* His interests include values clarification, teacher education, and curriculum reform.

Paul H. Hirst is Professor of Education at the University of Cambridge. He has written widely in philosophy of education and curriculum theory.

Among his publications are *The Logic of Education* (with R. S. Peters), *Moral Education in a Secular Society*, and *Knowledge and the Curriculum.*

Sidney Hook, a noted interpreter of Dewey's thought, is an eminent twentieth-century philosopher in his own right. Prior to his retirement, he was for many years chairman of the Department of Philosophy at New York University. His writings include *From Hegel to Marx, John Dewey: An Intellectual Portrait, Education for Modern Man, Political Power and Personal Freedom, The Quest for Being,* and *Philosophy and Public Policy.*

Lawrence Kohlberg is Professor of Education and Social Psychology and Director of the Center for Moral Education at Harvard University. He is the main spokesman for the cognitive-developmental approach to moral education and has written extensively on the subject. Among his writings are *Essays on Moral Development* (2 vols.), *Moral Stages,* and *The Meaning and Measurement of Moral Development.*

R. S. Peters is Emeritus Professor of Philosophy of Education at the University of London and a major figure in contemporary educational philosophy. His writings on education, ethics, and social philosophy include *Social Principles and the Democratic State, Authority, Responsibility and Education, Ethics and Education,* and *Essays on Educators.*

Louis E. Raths was credited with introducing the term "values clarification" at New York University during the 1950s. He was widely recognized as one of the pioneers in the values clarification movement. In addition to *Values and Teaching,* some of his influential works are *Teaching for Learning, Meeting the Needs of Children,* and *Values in the Classroom.*

Gilbert Ryle was Waynflete Professor of Metaphysical Philosophy at the University of Oxford and one of the most influential analytical philosophers of his time. Some of his most significant writings are *Philosophical Arguments, The Concept of Mind,* and *Dilemmas.* He was for many years the editor of *Mind.*

Israel Scheffler is Victor S. Thomas Professor of Education and Philosophy at Harvard University. He has made important contributions to the philosophy of science and has been one of the most influential American philosophers of education in recent years. His publications include *The Language of Education, The Anatomy of Inquiry, Conditions of Knowledge, Science and Subjectivity,* and *Of Human Potential.*

Betty A. Sichel is Professor of Philosophy of Education at Long Island University, C. W. Post Campus. Her scholarly interests are primarily in two areas, (1) theory of moral education and moral development and (2) classical Greek philosophy and how the concerns and ideas expressed in the Platonic corpus can still inform contemporary educational thought. Her articles have been published in *Educational Theory*, the *Journal of Moral Education*, *Curriculum Inquiry*, *Paideia*, the *Journal of Research and Development in Education*, and other journals.

Sidney B. Simon is Professor of Education at the University of Massachusetts. He is the co-author of several books relating to values clarification, including *Values and Teaching*, *Values Clarification: A Handbook of Practical Strategies for Teachers and Students*, and *Clarifying Values Through Subject Matter*.

John S. Stewart, former co-director of the Values Development Education Program, College of Education, Michigan State University, is president of Stewart Investments Seminars, Inc. in East Lansing, Michigan. He is the author of *The School as a Just Community* and numerous articles in educational journals. Dr. Stewart is particularly interested in individual rights and freedom within a context of community well-being.

Suggested Further Readings

BOOKS

Baier, Kurt. *The Moral Point of View*. Ithaca: Cornell University Press, 1963.

Bantock, G. H. *Education and Values*. London: Faber and Faber, 1965.

Barrow, Robin, *Moral Philosophy for Education*. London: Allen and Unwin, 1975.

Beck, Clive, Brian S. Crittenden, and Edmund V. Sullivan, eds. *Moral Education: Interdisciplinary Approaches*. New York: Newman Press, 1971.

Berkowitz, Marvin W., and Fritz Oser, eds. *Moral Education: Theory and Application*. Hillsdale, N. J.: Lawrence Erlbaum Associates, Inc., 1985.

Blackstone, William T., ed. *The Concept of Equality*. Minneapolis: Burgess Publishing Co., 1969.

Blackstone, William T., and George L. Newsome, eds. *Education and Ethics*. Athens: University of Georgia Press, 1969.

Brameld, Theodore, and Stanley Elam, eds. *Values in American Education*. Bloomington, Ind.: Phi Delta Kappa, 1964.

Brandt, Richard B. *Ethical Theory*. Englewood Cliffs, N. J.: Prentice-Hall, 1959.

Broudy, Harry. *Building a Philosophy of Education*. New York: Prentice-Hall, Inc., 1954.

Brown, L. M., ed. *Aims of Education*. New York: Teachers College Press, 1970.

Brubacher, John S. *Modern Philosophies of Education*. 4th ed. New York: McGraw-Hill Book Co., 1969.

Carter, Robert E. *Dimensions of Moral Education*. Toronto: University of Toronto Press, 1984.

Chazan, Barry. *Contemporary Approaches to Moral Education*. New York: Teachers College Press, 1985.

Chazan, Barry I., and Jonas F. Soltis, eds. *Moral Education*. New York: Teachers College Press, 1973.

Crittenden, Brian. *Form and Content in Moral Education*. Toronto: The Ontario Institute for Studies in Education, 1972.

Dewey, John. *Moral Principles in Education*. Boston: Houghton Mifflin, 1909.

Gilligan, Carol. *In A Different Voice: Psychological Theory and Women's Development*. Cambridge: Harvard University Press, 1982.

Hare, R. M. *Freedom and Reason*. Oxford: Clerendon Press, 1963.

Hare, R. M. *Moral Thinking: Its Levels, Method, and Point*. Oxford: Clarendon Press, 1981.

Harmin, Merrill, Howard Kirschenbaum, and Sidney Simon. *Clarifying Values through Subject Matter: Applications for the Classroom*. Minneapolis: Winston Press, 1973.

Jencks, Christopher, et al. *Inequality: A Reassessment of the Effect of Family and Schooling in America*. New York: Basic Books, 1972.

Kohlberg, Lawrence. *Essays on Moral Development*. Vol. 1. *The Philosophy of Moral Development*. San Francisco: Harper and Row, 1981.

Kohlberg, Lawrence. *Essays on Moral Development*. Vol. 2. *The Psychology of Moral Development*. San Francisco: Harper and Row, 1984.

Kohlberg, Lawrence. *The Meaning and Measurement of Moral Development*. Worcester, Mass.: Clark University, 1981.

Kohlberg, Lawrence, Charles Levine, and Alexandra Hewer. *Moral Stages: A Current Formulation and a Response to Critics*. New York: Karger, 1983.

Lerner, Max. *Values in Education: Notes Toward a Values Philosophy*. Bloomington, Ind.: Phi Delta Kappa, 1976.

MacIntyre, Alasdair C. *After Virtue: A Study in Moral Theory*. Notre Dame, Ind.: University of Notre Dame Press, 1981.

McClelland, David C., ed. *Education for Values*. New York: Irvington Publishers, Inc., 1982.

Meyer, John R., Brian Burnham, and John Cholvat, eds. *Values Education*. Waterloo, Ontario: Wilfrid Laurier University Press, 1975.

Metcalf, Lawrence E., ed. *Values Education*. Forty-first Yearbook of the National Council for the Social Studies. Washington, D. C.: NCSS, 1971.

Munsey, Brenda, ed. *Moral Development, Moral Education, and Kohlberg: Basic Issues in Philosophy, Psychology, Religion, and Education*. Birmingham, Ala.: Religious Education Press, 1980.

Nash, Paul. *Authority and Freedom in Education*. New York: Wiley and Sons, Inc., 1966.

Noddings, Nel. *Caring, A Feminine Approach to Ethics and Moral Education*. Berkeley: University of California Press, 1984.

Nowell-Smith, P. H. *Ethics*. London: Penguin Books, 1954.

Nozick, Robert. *Anarchy, State and Utopia*. New York: Basic Books, Inc., 1974.

Perry, Ralph Barton. *Realms of Value*. Cambridge, Mass.: Harvard University Press, 1954.

Peters, R. S. *Authority, Responsibility, and Education*. London: George Allen & Unwin, Ltd., 1959.

Peters, R. S. *Moral Development and Moral Education*. London: George Allen & Unwin, Ltd., 1981.

Phenix, Philip H. *Philosophy of Education*. New York: Henry Holt and Co., Inc., 1958.

Phillips, D. C. *Theories, Values and Education*. Melbourne: Melbourne University Press, 1971.

Piaget, Jean. *The Moral Judgment of the Child*. Translated by Marjorie Gabain. New York: The Free Press, 1965.

Purpel, David, and Kevin Ryan, eds. *Moral Education . . . It Comes with the Territory*. Berkeley, Cal.: McCutchan, 1976.

Rawls, John. *A Theory of Justice*. Cambridge, Mass.: Harvard University Press, 1971.

Rich, John Martin. *Education and Human Values*. Reading, Mass.: Addison-Wesley, 1968.

Rich, John Martin, and Joseph L. DeVitis. *Theories of Moral Development*. Springfield, Ill.: Charles C. Thomas, 1985.

Sellars, Wilfrid, and John Hospers, eds. *Readings in Ethical Theory*. New York: Appleton-Century-Crofts, 1952.

Shaver, James P. *Values and Schooling*. Logan, Utah: Utah State University, 1972.

Shaver, James P., and William Strong. *Facing Value Decisions: Rationale-building for Teachers*. 2d ed. New York: Teachers College Press, 1982.

Simon, Sidney B., Leland W. Howe, and Howard Kirschenbaum. *Values Clarification: A Handbook of Practical Strategies for Teachers and Students*. 2d ed. New York: Hart, 1978.

Simon, Sidney B., Howard Kirschenbaum, eds. *Readings in Values Clarification*. Minneapolis: Winston Press, 1973.

Sloan, Douglas, ed. *Education and Values*. New York: Teachers College Press, 1980.

Smith, Philip G. *Philosophy of Education*. New York: Harper and Row, 1964.

Smith, Philip G., ed. *Theories of Value and Problems of Education*. Urbana, Ill.: University of Illinois Press, 1970.

Smith, Philip L. *The Problem of Values in Educational Thought*. Ames: Iowa State University Press, 1982.

Snook, Ivan, and Colin Lankshear. *Education and Rights*. Melbourne: Melbourne University Press, 1979.

Straughan, Roger. *Can We Teach Children to be Good?* London: Allen and Unwin, 1982.

Strike, Kenneth A. *Educational Policy and the Just Society*. Champaign-Urbana, Ill.: The University of Illinois Press, 1982.

Strike, Kenneth A. *Liberty and Learning*. New York: St. Martin's Press, 1982.

Strike, Kenneth A., and Jonas F. Soltis. *The Ethics of Teaching*. New York: Teachers College Press, 1985.

Toulmin, Stephen. *An Examination of the Place of Reason in Ethics*. Cambridge: Cambridge University Press, 1950.

Vandenberg, Donald. *Human Rights in Education*. New York: Philosophical Library, 1983.

Volkmor, Cara B., Anne Langstaff Pasanella, and Louis E. Raths. *Values in the Classroom*. Columbus, Ohio: Charles E. Merrill, 1977.

Wilson, John, Norman Williams, and Barry Sugarman. *Introduction to Moral Education*. London: Penguin Books, 1967.

Wringe, C. A. *Children's Rights: A Philosophical Study*. Boston: Routledge & Kegan Paul, 1981.

ARTICLES

Aiken, Henry D. "Moral Philosophy and Education." *Harvard Educational Review* 25 (Winter 1955): 39–59.

Alston, William P. "Comments on Kohlberg's 'From Is to Ought.'" In *Cognitive Development and Epistemology*. Edited by Theodore Mischel. New York: Academic Press, 1971: 269–284.

Arnstine, Donald G. "Some Problems in Teaching Values." *Educational Theory* 11 (July 1961): 158–167.

Atkinson, R. F. "Instruction and Indoctrination." In *Philosophical Analysis and Education*. Edited by Reginald D. Archambault. New York: The Humanities Press, 1965.

Bandman, Bertram. "Is There a Right to an Education?" In *Philosophy of Education 1977*, Proceedings of the Thirty-third Annual Meeting of the Philosophy of Education Society, 1977: 287–297.

Baron, Marcia. "The Ethics of Duty/Ethics of Virtue Debate and Its Relevance to Educational Theory." *Educational Theory* 35 (Spring 1985): 135–149.

Barrow, Robin. "Socrates was a Human Being: A Plea for Transcultural Moral Education." *The Journal of Moral Education* 15 (January 1986): 50–57.

Beck, Clive M. "The Reflective Approach to Values Education." In *Philosophy and Education*. Eightieth Yearbook of the National Society for the Study of Education. Chicago: University of Chicago Press, 1981.

Benedict-Gill, Diane. "A Subject Matter Description of Moral Education." *Educational Theory* 25 (Spring 1975): 103–115.

Berlak, Harold. "Values, Goals, Public Policy and Educational Evaluation." *Review of Educational Research* 40 (April 1970): 261–278.

Berlin, Isaiah. "Equality as an Ideal." In *Justice and Social Policy*. Edited by Frederick A. Olafson. Englewood Cliffs, N. J.: Prentice-Hall, 1961. (Originally published as "Equality" in the *Proceedings of the Aristotelian Society*, vol. 56, 1955–56. London: Harrison and Sons, 1956.)

Berson, Robert. "The Educational Situation and the Realm of Values." *Educational Theory* 25 (Spring 1975): 125–130.

Beversluis, Eric H. "Benefit Rights in Education: An Entitlement View." In *Philosophy of Education 1984*, Proceedings of the Fortieth Annual Meeting of the Philosophy of Education Society, 1984, pp. 381–390.

Beversluis, Eric H. "The Dilemma of Values Clarification." In *Philosophy of Education* 1978, Proceedings of the Thirty-fourth Annual meeting of the Philosophy of Education Society, 1978: 417–427.

Beyer, Landon E., and George H. Wood. "Critical Inquiry and Moral Action in Education." *Educational Theory 36* (Winter 1986): 1–14.

Bibby, Martin. "Aims and Rights." *Educational Philosophy and Theory* 17 (October 1985): 1–11.

Boyd, Dwight, and Deanne Bogdan. "Something Clarified, Nothing of Value: A Rhetorical Critique of Values Clarification." *Educational Theory* 34 (Summer 1984): 287–300.

Butler, J. Donald. "The Role of Value Theory in Education." *Educational Theory* 4 (January 1954): 69–77, 86.

Caplan, Arthur L. "'Ethics' and 'Values' in Education: Are the Concepts Distinct and Does It Make a Difference?" *Educational Theory* 29 (Summer 1979): 245–253.

Claydon, L. F. "Teaching and Commitment to Values." *Educational Philosophy and Theory* 5 (March 1973): 1–8.

Clayton, A. Stafford. "Education and Some Moves Toward a Value Methodology." *Educational Theory* 19 (Spring 1969): 198–210.

Coleman, James S. "The Concept of Equality of Educational Opportunity." *Harvard Educational Review* 38 (Winter 1968): 7–22.

Cooper, Jane Wilcox. "An Analysis of the Question of Values and Evaluation in Educational Philosophy." *Educational Theory* 4 (January 1954): 4–15, 26.

Counts, George S. "Should the Teacher Always Be Neutral." *Phi Delta Kappan* 51 (December 1969): 186–189.

Craig, Robert P. "Form, Content and Justice in Moral Reasoning." *Educational Theory* 26 (Spring 1976): 154–157.

Crittenden, Brian. "Aims, Intentions and Purposes in Teaching and Educating." *Educational Theory* 24 (Winter 1974): 46–51.

Crittenden, Brian S. "What Is Educational Value?" *The Journal of Value Inquiry* 2 (Winter 1968): 235–248.
Delattre, Edwin J., and William J. Bennett. "Where the Value Movement Goes Wrong." *Change* 11 (February 1979): 38–43.
Dewey, John. "Teaching Ethics in the High School." *Educational Theory* 17 (July 1967): 222–226, 247.
Elliot, Robert. "Curriculum, Morality, and Theories About Value." *Educational Philosophy and Theory* 14 (October 1982): 15–28.
Elvin, Lionel. "Individuality and Education." *British Journal of Educational Studies* 28 (June 1980): 87–99.
Engel, David E. "Some Issues in Teaching Values." *Religious Education* 55 (January-February 1970): 9–13.
Ennis, Robert H. "Equality of Educational Opportunity." *Educational Theory* 26 (Winter 1976): 3–18.
Frankena, William. "Toward a Philosophy of Moral Education." *Harvard Educational Review* 28 (Fall 1958): 300–313.
Gall, Morris. "Some Value Problems of the Classroom Teacher." *Educational Theory* 4 (October 1954): 297–299.
Giarelli, James M. "Lawrence Kohlberg and G. E. Moore on the Naturalistic Fallacy." *Educational Theory* 26 (Fall 1976): 348–354.
Gordon, David. "The Immorality of the Hidden Curriculum." *Journal of Moral Education* 10 (October 1980): 3–8.
Green, Thomas F. "Equality of Educational Opportunity: The Durable Injustice." In *Philosophy of Education,* Proceedings of the Twenty-seventh Annual Meeting of the Philosophy of Education Society, 1971: 121–143.
Green, Thomas F. "The Formation of Conscience in an Age of Technology." *American Journal of Education* 94 (November 1985): 1–32.
Haldane, John J. "Concept-Formation and Value Education." *Educational Philosophy and Theory* 16 (October 1984): 22–27.
Hardie, C. D. "The Idea of Value and the Theory of Education." *Educational* Theory 7 (July 1957): 196–199.
Heslep, Robert D. "Rawls and Meritocratic Education." In *Philosophy of Education 1980,* Proceedings of the Thirty-sixth Annual Meeting of the Philosophy of Education Society, 1980: 190–198.
Johnson, Conrad D. "The Morally Educated Person in a Pluralistic Society." *Educational Theory* 31 (Summer-Fall 1981): 237–250.
Kirschenbaum, Howard, Merrill Harmin, Leland Howe, and Sidney B. Simon. "In Defense of Values Clarification." *Phi Delta Kappan* 58 (June 1977): 743–744.
Kleinberger, Aharon Fritz. "Reflections on Equality in Education." *Studies in Philosophy and Education* 5 (Summer 1967): 293–340.
Kohlberg, Lawrence. "From Is to Ought: How to Commit the Naturalistic

Fallacy and Get Away with It in the Study of Moral Development." In *Cognitive Development and Epistemology*. Edited by Theodore Mischel. New York: Academic Press, 1971: 155–235.

Komisar, Paul and Jerrold Coombs. "Too Much Equality." *Studies in Philosophy and Education* 4 (Fall 1965): 263–271.

Leming, James S. "Curricular Effectiveness in Moral/Values Education." *Journal of Moral Education* 10 (May 1981): 147–164.

Lepley, Ray. "The Current Status of Value Theory." *Educational Theory* 4 (April 1954): 158–165.

Lieberman, Myron. "Equality of Educational Opportunity." *Harvard Educational Review* 29 (Summer 1959): 167–183.

Liston, Daniel P. "On Facts and Values." *Educational Theory* 36 (Spring 1986): 137–152.

Lockwood, Alan L. "A Critical View of Values Clarification." *Teachers College Record* 77 (September 1975): 35–50.

Macmillan, C. J. B. "Equality and Sameness." *Studies in Philosophy and Education* 3 (Winter 1964–65): 320–332.

Magsino, Romulo F. "Freedom and Rights in Schools: Towards Just Entitlements for the Young." *Educational Theory* 29 (Summer 1979): 171–185.

Montifiore, Alan. "Moral Philosophy and the Teaching of Morality." *Harvard Educational Review* 35 (Fall 1965): 435–449.

Olafson, Frederick A. "Rights and Duties in Education." In *Educational Judgments*. Edited by James F. Coyle. London: Routledge & Kegan Paul, 1973: 173–195.

Oppewal, Donald. "Democracy and Democratic Values: Their Status in Educational Theory." *Educational Theory* 9 (July 1959): 156–164.

Peters, R. S. "Democratic Values and Educational Aims." *Teachers College Record* 80 (February 1979): 463–482.

Peters, R. S. "Moral Development: A Plea for Pluralism." In *Cognitive Development and Epistemology*. Edited by Theodore Mischel. New York: Academic Press, 1971: 237–267.

Peters, R. S. "Reason and Habit: The Paradox of Moral Education." In *Philosophy and Education*. 2d ed. Edited by Israel Scheffler. Boston: Allyn and Bacon, Inc., 1966. (Originally published in *Moral Education in a Changing Society*. Edited by W. R. Niblett. London: Faber and Faber, 1963.)

Powell, J. P. "On Justifying a Broad Educational Curriculum." *Educational Philosophy and Theory* 2 (March 1970): 53–61.

Prakash, Madhu S. "In Pursuit of Wholeness: Moral Development, the Ethics of Care and the Virtue of Philia." In *Philosophy of Education 1984*, Proceedings of the Fortieth Annual Meeting of the Philosophy of Education Society, 1984: 63–74.

Proefriedt, William. "Power, Pluralism, and the Teaching of Values: The Educational Marketplace." *Teachers College Record* 86 (Summer 1985): 513–537.

Rich, John Martin. "Moral Education and the Emotions." *Journal of Moral Education* 9 (January 1980): 81–87.

Scriven, Michael. "Cognitive Moral Education." *Phi Delta Kappan* 56 (June 1975): 689–694.

Sichel, Betty A. "Can Kohlberg Respond to Critics." *Educational Theory* 26 (Fall 1976): 337–347, 394.

Siegel, Harvey. "Kohlberg, Moral Adequacy, and the Justification of Educational Interventions." *Educational Theory* 31 (Summer-Fall 1981): 275–284.

Simon, Sidney B., and Polly de Sherbinin. "Values Clarification: It Can Start Gently and Grow Deep." *Phi Delta Kappan* 56 (June 1975): 679–683.

Simpson, Evan. "A Values-Clarification Retrospective." *Educational Theory* 36 (Summer 1986): 271–287.

Smith, Philip G. "Knowledge and Values." *Educational Theory* 26 (Winter 1976): 29–39.

Sommers, Cristina Hoff. "Ethics Without Virtue." *American Scholar* 53 (Summer 1984): 381–389.

Strike, Kenneth A. "Fairness and Ability Grouping." *Educational Theory* 33 (Summer-Fall 1983): 125–134.

Strike, Kenneth A. "Freedom, Autonomy, and Teaching." *Educational Theory* 22 (Summer 1972): 262–277.

Suttle, Bruce B. "Moral Education Versus Values Clarification." *Journal of Educational Thought* 16 (April 1982): 35–41.

Tyack, David B., and Thomas James. "Moral Majorities and the School Curriculum: Historical Perspectives on the Legalization of Virtue." *Teachers College Record* 86 (Summer 1985): 513–537.

Urmson, J. O. "On Grading." *Mind* 59 (April 1950): 145–169.

Vandenberg, Donald. "Human Dignity, Three Human Rights and Pedagogy." *Educational Theory* 36 (Winter 1986): 33–42.

Waks, Joseph. "Education and Meta-Ethics." *Studies in Philosophy and Education* 6 (Spring 1969): 351–359.

Wilson, John. "Moral Education: Retrospect and Prospect." *Journal of Moral Education* 9 (October 1979): 3–9.

Worsfold, Victor. "A Philosophical Justification for Children's Rights." *Harvard Educational Review* 44 (February 1974): 142–157.

Index